Kitchen chemistry

Written by Ted Lister
in collaboration with
Heston Blumenthal

Kitchen chemistry

Written by Ted Lister in collaboration with Heston Blumenthal

Edited by Colin Osborne and Emma Kemp

Designed by Russel Spinks and Alberto Arias at www.russel-spinks.co.uk

Photographs by Russel Spinks

Published and distributed by Royal Society of Chemistry

Printed by Royal Society of Chemistry

For further information on other educational activities undertaken by the Royal Society of Chemistry contact:
Email: *education@rsc.org*
Telephone: 020 7440 3344

Education Department
Royal Society of Chemistry
Burlington House
Piccadilly
London
W1J 0BA

Information on other Royal Society of Chemistry activities can be found on its websites:
www.rsc.org
www.chemsoc.org
www.chemsoc.org/LearnNet contains resources for teachers and students from around the world.

ISBN 0-85404-389-6

British Library Cataloguing in Publication Data.

A catalogue for this book is available from the British Library.

Foreword

Chemistry is all around us and affects every aspect of our daily lives, but all too often we overlook the beneficial impact of chemical sciences. This resource sets out some chemistry relevant to the school and college curriculum that is used daily in kitchens both in homes and restaurants, and which makes the food we eat more pleasurable.

The Royal Society of Chemistry is pleased that some chefs are bringing a scientific approach to their kitchen skills and hopes that this work will lead to an increased awareness of the role of science in general, and chemistry in particular, in preparing the food we eat.

Dr Simon Campbell FRSC FRS

President, Royal Society of Chemistry

As this book went to press, Heston Blumenthal's restaurant, The Fat Duck, was named best restaurant in the world by *Restaurant* magazine.

ELGA

Introduction

One of the most exciting things that has happened at my restaurant, The Fat Duck, recently is the Royal Society of Chemistry producing this resource for schools – *Kitchen chemistry*. It is based on taking a scientific approach to cooking – an activity that has traditionally been regarded as an art, rather than a science. Topics range from the simple (what is the role of salt in cooking vegetables?) to the complex (separating volatile flavour components in foods by gas chromatography mass spectrometry), to the 'just for fun' (breaking the world record for ice cream making by using liquid nitrogen as a coolant). What the RSC has done is to provide flexible material that teachers can 'dip into' that relates the chemistry that goes on in the home or restaurant kitchen to that which students learn about in the school curriculum.

Kitchen chemistry makes chemistry more accessible because it brings together scientific theory and everyday practicality. After all, we all know something about cooking even though we may not do it very often, and children are no different. When I left school I had no scientific background whatsoever. I have taught myself slowly and with much difficulty, so this new initiative is music to my ears. I just wish it had happened a few years earlier.

Heston Blumenthal

Chef and proprietor of The Fat Duck

Mango and Douglas Fir Pureé: bavarois of lychee and mango, blackcurrant and green pepercorn jelly

Interior of The Fat Duck restaurant

Cauliflower risotto, carpaccio of cauliflower, chocolate jelly

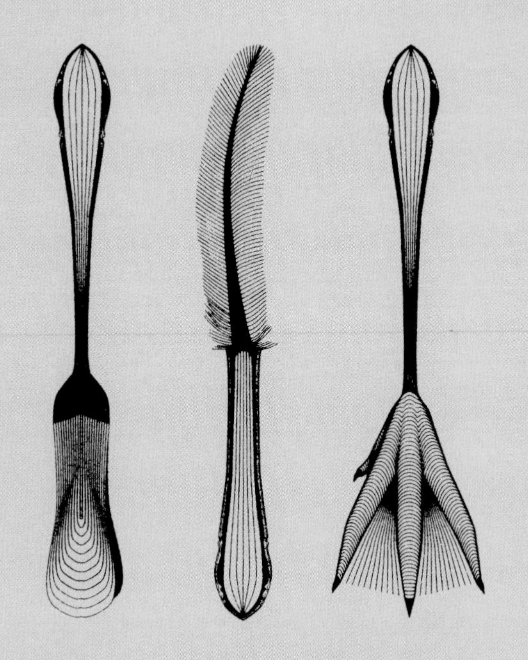

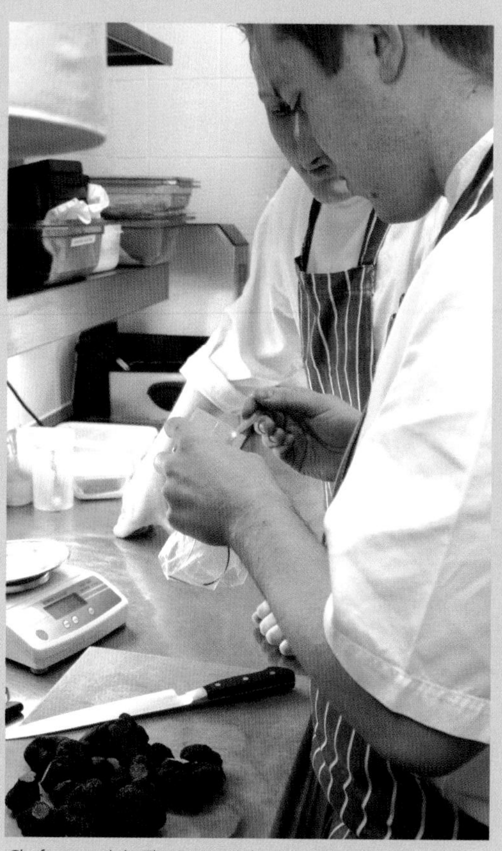

Chefs at work in The Fat Duck kitchen

Rhubarb galette, rhubarb sorbet

Heston Blumenthal with author Ted Lister

Contents

How to use this material

All food is, of course, made of chemicals, and cooking can be thought of as a series of chemical reactions in which changes occur to some of these chemicals. The aims of cooking are several:

- to kill microorganisms and denature enzymes that might bring about undesirable changes in food
- to maintain or enhance the nutritional value of the food
- to improve the texture of the food
- to improve the appearance of the food
- to improve the flavour of the food
- to improve the aroma of the food.

The material presented here looks at various aspects of the chemistry of food and the cooking process. It consists of activities of a variety of types – class practical, demonstration experiments, reading comprehension and paper-based activities – at a variety of levels from primary to post-16. The index table will allow users to select an activity of an appropriate topic, type and level. Each activity deals with an aspect of the chemistry of food and/or cooking. Although the chemistry of food and cooking is not directly part of most curricula, it can often be used to show familiar chemistry in a context that may be stimulating for many students. The material also allows teachers to reinforce the idea that everything is made of chemicals and that there is no difference between 'man-made' and 'natural' chemicals. In particular there are a number of activities on which experimental investigations can be based – helping to cover this part of the English National Curriculum and equivalents. Some of the paper-based or comprehension activities could be used as revision lessons or in the case of teacher absence.

The material is presented as teacher's notes and student worksheets. The worksheets are available on the CDROM accompanying this book or may be downloaded from ***http://www.chemsoc.org/kitchenchemistry*** as colour or black and white pdf files, or as Microsoft® Office Word documents (which can be edited by the teacher if required). Also included on the CDROM and website are video clips related to some of the material. These may be played to start off a lesson or stimulate discussion. However, all the lessons can be tackled without the use of the video clips for those who prefer not to use them. In every case, material is given that the teacher can use to start the lesson by discussion.

The video clips are taken from the Discovery Channel TV series, *Kitchen Chemistry*, featuring Heston Blumenthal. Heston is a chef and proprietor of The Fat Duck, a Michelin three-star restaurant in Bray, Berkshire. He is noted for his scientific approach to food and cooking and for the fact that he will not take for granted the accepted wisdom without scientifically investigating it for himself. He also makes use of scientific equipment in the kitchens of the Fat Duck – temperature probes, desiccators and reflux apparatus, for example.

The index of topics (see page xiii) will allow a quick overview of the material available according to topic and age group.

Many of the activities involve practical work with food and a number of them involve tasting. It is important to stress that tasting in an exceptional activity in a science laboratory and teachers should take appropriate precautions with regard to hygiene. Reproduced below are recommendations from CLEAPSS (The Consortium of Local Authorities for Provision of Science Services) about experimental work with food.

Experiments with food

The following advice has been adapted (with permission) from the *CLEAPSS School Science Service Laboratory Handbook.*

When experiments with food are carried out, it is worth arranging for the class to be transferred to the Home Economics/Food Technology Department for the session, especially if the food is to be tasted. This minimises risks and also reinforces the special nature of laboratories in students' minds. If a laboratory has to be used for tasting activities it should be made very clear to students that tasting activities are exceptional, and that normally, eating and drinking in the laboratory are not permitted.

Particular attention must be paid to hygiene for tasting investigations. For example all bench surfaces should be cleaned and preferably disinfected so that students do not inadvertently pick up spilled chemicals on their hands. The use of plastic sheets as for microbiology experiments may be considered. All equipment must be scrupulously clean and where a student tastes food with, for example, a teaspoon, this must be adequately sterilised before reuse by another student. The use of disposable plates, cups and spoons is preferable. If normal laboratory glassware is to be used, it is a good idea to have separate stocks that are reserved solely for tasting investigations. Even in investigations where food is not intended to be tasted, there may be a temptation for students to try to taste it. They should be warned against this.

More detailed advice and model risk assessments on experiments with food can be found on the CDROM *CLEAPSS Design & Technology Publications*, Uxbridge: CLEAPSS School Science Service.

Blindfold tasting of cooked beans is conducted in a Food Technology room using disposable plates with a disposable plastic spoon for each student

Heston Blumenthal outside The Fat Duck restaurant

Interior of The Fat Duck restaurant

Chefs at work in The Fat Duck kitchen

Roast Foie Gras, chamomile, almond, cherry and Amaretto jelly

Guide to the layout of teacher's notes

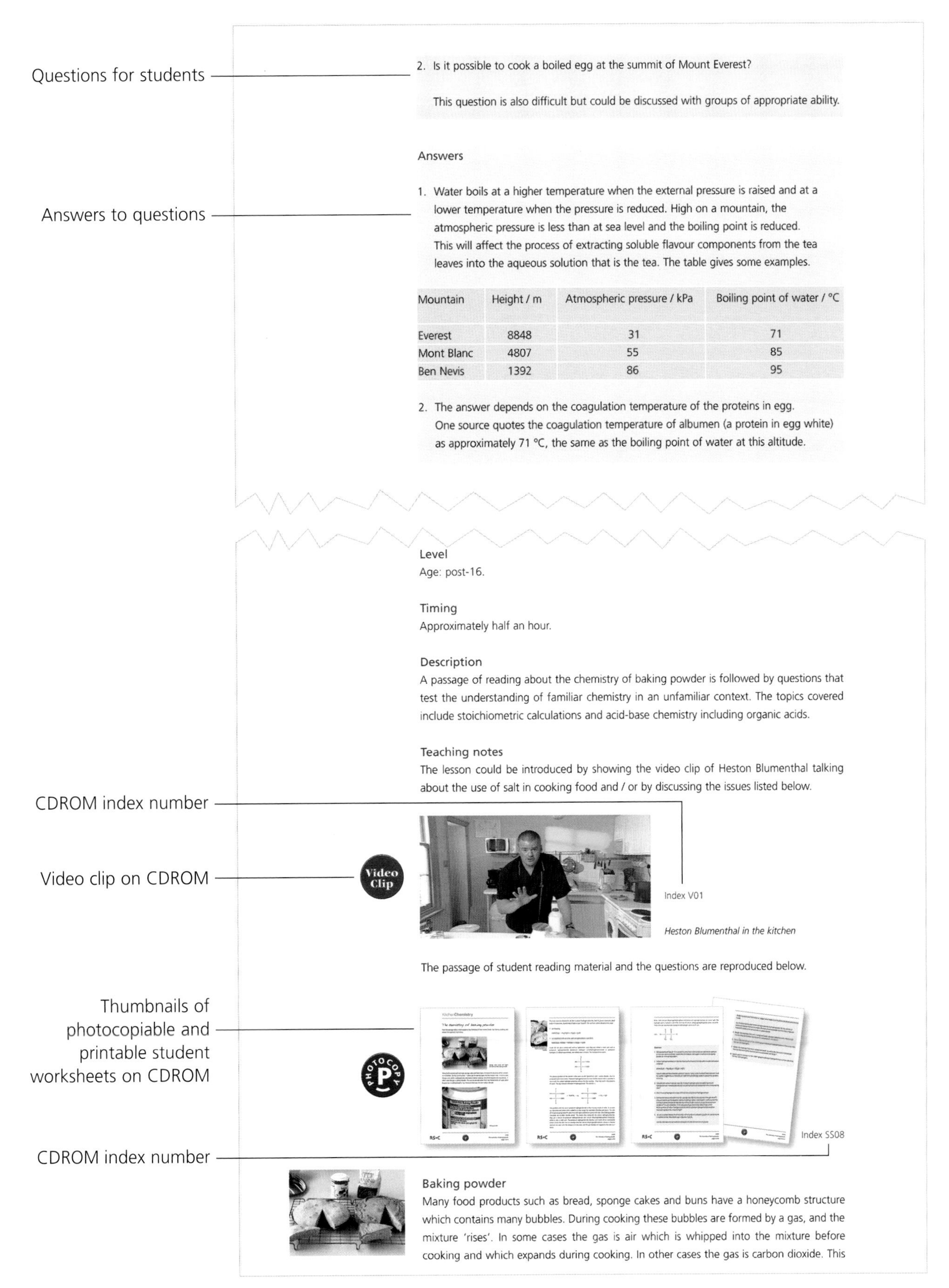

2. Is it possible to cook a boiled egg at the summit of Mount Everest?

This question is also difficult but could be discussed with groups of appropriate ability.

Answers

1. Water boils at a higher temperature when the external pressure is raised and at a lower temperature when the pressure is reduced. High on a mountain, the atmospheric pressure is less than at sea level and the boiling point is reduced. This will affect the process of extracting soluble flavour components from the tea leaves into the aqueous solution that is the tea. The table gives some examples.

Mountain	Height / m	Atmospheric pressure / kPa	Boiling point of water / °C
Everest	8848	31	71
Mont Blanc	4807	55	85
Ben Nevis	1392	86	95

2. The answer depends on the coagulation temperature of the proteins in egg. One source quotes the coagulation temperature of albumen (a protein in egg white) as approximately 71 °C, the same as the boiling point of water at this altitude.

Level
Age: post-16.

Timing
Approximately half an hour.

Description
A passage of reading about the chemistry of baking powder is followed by questions that test the understanding of familiar chemistry in an unfamiliar context. The topics covered include stoichiometric calculations and acid-base chemistry including organic acids.

Teaching notes
The lesson could be introduced by showing the video clip of Heston Blumenthal talking about the use of salt in cooking food and / or by discussing the issues listed below.

Index V01

Heston Blumenthal in the kitchen

The passage of student reading material and the questions are reproduced below.

Index SS08

Baking powder
Many food products such as bread, sponge cakes and buns have a honeycomb structure which contains many bubbles. During cooking these bubbles are formed by a gas, and the mixture 'rises'. In some cases the gas is air which is whipped into the mixture before cooking and which expands during cooking. In other cases the gas is carbon dioxide. This

Extracts of sample pages from this book

Blumenthal hits peak of haute cuisine

For years Britain has been cast as the poor relation when it comes to food, but last night Heston Blumenthal's restaurant The Fat Duck, in Bray, Berkshire, was crowned the best in the world.

A panel of 600 international chefs and critics voted for the man whose dishes sound like a mistake – smoked bacon and egg ice cream or green tea and lime mousse dipped in liquid nitrogen.

The Times, 19 April 2005

There is no doubt Heston Blumenthal is the most original and remarkable chef this country has ever produced

The Fat Duck, the pioneering British restaurant that introduced the world to delicacies such as sardine on toast sorbet and bacon and egg ice cream, has been declared the world's best place to eat.

Chef Heston Blumenthal's restaurant in the Berkshire village of Bray topped a list of the world's 50 best restaurants which was unveiled in London last night.

The Guardian, 19 April 2005

Heston Blumenthal's well equipped kitchen laboratory at The Fat Duck restaurant

General introduction

Heston Blumenthal

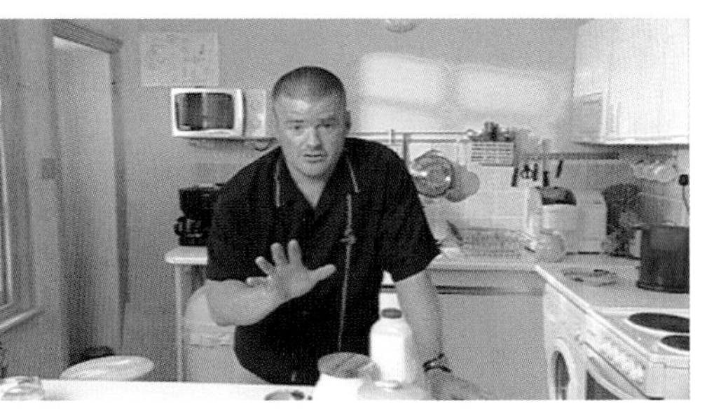

Index V01

Heston Blumenthal in the kitchen

This clip could be used to start a general discussion about food and the fact that it is, of course, made up of chemicals and that chemicals in food are not just undesirable additives or impurities. It could also be used at the start of a chemistry course when asking students what chemicals are and when discussing how the media sometimes portray all chemicals as dangerous.

Heston Blumenthal is a chef and owner of The Fat Duck restaurant in Bray, Berkshire. In 2004 the restaurant was awarded its third Michelin star – a rare accolade. Heston is particularly noted for his scientific approach to cooking. This shows itself in a number of ways:

- his willingness to ask questions rather than take for granted the accepted wisdom
- his testing of theories by experiment
- his use of laboratory equipment - temperature probes, reflux apparatus, *etc* in his cooking.

At the time of writing, Heston has just been given funding to take on a PhD student in conjunction with Nottingham University to study 'hydrocolloid systems with unusual mechanical and thermal properties'. This will have applications in improving ice cream, gravies and jellies, among other things. The student will also study 'novel flavour-release mechanisms using nanotechnology' to determine different ways in which food gives off its flavour.

For more information on Heston and The Fat Duck, see
http://www.fatduck.co.uk/intro.html (*accessed January 2005*)

Acknowledgements
The Royal Society of Chemistry would like to thank the following people and organisations for their help in producing this material.
Heston Blumenthal and his staff at The Fat Duck restaurant, Bray, Berkshire
The Discovery Channel
The CLEAPSS School Science Service
The Maplesden Noakes School, Maidstone, Kent
Highsted Grammar School, Sittingbourne, Kent.

Selected bibliography

T. Coultate and J. Davies, *Food, the definitive guide*, Cambridge: Royal Society of Chemistry, 1994

R. L. Wolke, *What Einstein Told His Cook*, New York: W. W. Norton & Co., 2002

H. Blumenthal, *Family Food*, London: Penguin, 2002

P. Barham, *The Science of Cooking*, Berlin: Springer, 2001

B. A. Fox and A. G. Cameron, *Food Science a Chemical Approach*, Sevenoaks: Hodder and Stoughton, 1982

T. Coultate, *Food, the Chemistry of Its Components*, Cambridge: Royal Society of Chemistry, 1992

E. N. Ramsden, *Biochemistry and Food Science*, Cheltenham: Stanley Thornes, 1995

Food Science, a Special Study, London: Longman, 1974

Revised Food Science, a Special Study, London: Longman, 1984

Note. Information on the Nuffield Chemistry Special Study materials is available at *http://www.chemistry-react.org* (*accessed January 2005*)

The Most Useful Science Harleston: Thorburn Kirkpatrick, 1989

Heston Blumenthal writes an occasional column in *The Guardian* Weekend Colour Supplement.

A useful list of food-based experiments and links to the websites of many relevant organisations can be found at *www.foodtechcareers.org* (*accessed January 2005*).

Index of topics

Topic title	*Type of activity*	*Age 5–11*	*Age 11–16*	*Post-16*	*Downloadable resources*	*Video clip(s)*
The use of salt in cooking (1)	CP	?	✔	?	Photocopy, Video Clip	V01
The use of salt in cooking (2)	CP	✗	?	✔	Photocopy, Video Clip	V01
By how much does salt increase the boiling point of water?	CP	✗	✔	?	Photocopy, Video Clip	V01
Is all salt the same?	W	?	✔	✔	Power Point, Photocopy	-
'Low sodium' salt substitutes	CP	✗	?	✔	Photocopy	-
What affects the colour and texture of cooked vegetables?	CP/W	✗	?	✔	Photocopy, Video Clip	V01, V14
Should beans be cooked with the lid on or off?	CP	✔	?	✗	Photocopy, Video Clip	V01
The chemistry of baking powder	W	✗	✗	✔	Photocopy, Video Clip	V01
The structure of ice and water	D/W	✗	✗	✔	Photocopy	-
Why do pans stick?	W	✗	✗	✔	Photocopy	-
Enzymes & jellies	D	✔	?	✗	Photocopy, Video Clip	V02, V03
Chemistry of flavour	W	✗	✗	✔	Photocopy, Video Clip	V04, V05, V07, V08, V06
Chemical changes during cooking	W	✗	✗	✔	Photocopy, Video Clip	V09, V10
The science of ice cream	D/W	✔	✔	✔	Video Clip	V11, V12, V13
'Asparagus pee'	W	✗	✗	✔	Photocopy, Video Clip	V07
How hot are chilli peppers?	W	✗	✗	✔	Photocopy	-

Key to index of topics

✔ suitable for this age group

? some of the material is suitable or could be adapted for this age group

✗ unsuitable for this age group

W written exercise, *eg* reading and comprehension, answering questions

CP class practical activity

D teacher demonstration

photocopiable and printable worksheet

video clip

Chime file

PowerPoint® presentation

CDROM instructions and system requirements

The CDROM is compatible with Windows 98/NT/2000/ME/XP.
Recommended minimum specification: 64Mb RAM.

Insert the CDROM into the CDROM drive. Your PC should run the CDROM automatically. If it does not, double click your CDROM drive, using Windows Explorer or My Computer.

You will need :

- Web Browser – the content has been optimised for Internet Explorer 6
- Adobe Acrobat® PDF Reader – to use the PDF files in the resources section
- Microsoft® Office Word – to use the documents in the resources section
- Microsoft PowerPoint® – to view the PowerPoint® presentation
- Chime plug-in for your browser – to view rotatable models of molecules.

The CDROM will prompt you to install the Adobe Acrobat® PDF Reader and/or the Chime plug-in if required. It is the responsibility of the user to have MS Word and PowerPoint ® installed on their system before running this CDROM.

The CDROM licence allows the files on the CDROM to be downloaded and to be accessible over a network. The RSC will not offer support or guidance on how best to network the files.

Disclaimer

This CDROM has been thoroughly checked for errors and viruses. The RSC cannot accept liability for any damage to your computer system or data which occurs while using this CDROM or the software contained on it. If you do not agree with these conditions, you should not use the CDROM.

Format of video clips

The 16 short video clips available on the CDROM which accompanies this book are all approximately 40 Mb mpg files which can be played using a variety of media players and which should be suitable for full screen playback.

These video clips are also available in two file sizes from the RSC website for playback using Windows Media® Player:

1. **2 Mb** Windows Media® Video (wmv) files. These will be of too low a quality for playing at full screen size and will therefore be unsuitable for class use. They are, however, quick to download (approximately 30 seconds on a broadband connection and 5 minutes via a 56 k modem) and should enable teachers to decide whether they will find the clip useful and therefore worth downloading in a higher resolution format.
2. **15 Mb** wmv files. These should be suitable for playing at full screen but will take longer to download – about 4 min on a broadband connection and about 40 minutes via a 56 k modem.

List of video clips

V01 Using salt in cooking

This is probably the key clip. In it Heston Blumenthal describes a 'defining moment' in his career as a chef when he began to use science in his cooking and started to carry out experiments rather than follow the received wisdom. This occurred when he first thought to ask why salt is always added to the water when cooking green vegetables.

V02 What are jellies?

Heston Blumenthal discusses gelatine as a setting agent for various types of jelly.

V03 Pineapple jelly

Gelatine-based jellies will not set if they are made with pineapple, which contains an enzyme that denatures the protein in gelatine. Heston Blumenthal shows the effect of this enzyme on the proteins in his own mouth and shows a way to avoid the non-setting problem using chillies.

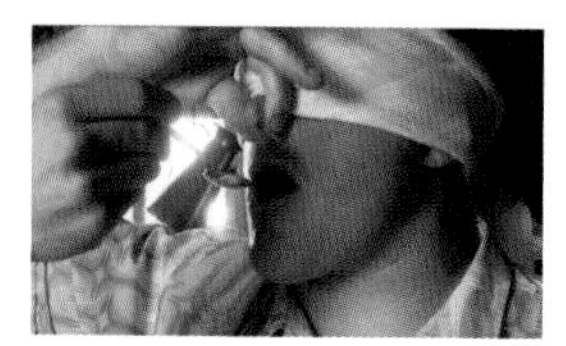

V04 What is flavour?

This reinforces the idea that flavour is taste plus aroma and shows tasting experiments in which a blindfolded taster holds his nose and becomes unable to identify flavour.

V05 Flavour, taste and aroma

This shows professional tasters also unable to identify flavour when their noses are blocked.

V06 The taste of food

Heston Blumenthal discusses with a scientist colleague, Peter Barham from Bristol University, the five taste sensations (sweet, sour, bitter, salty and the recently-accepted umami) and how our responses to them have evolved.

V07 Cooking asparagus

This explains that asparagus should be cooked in butter rather than water because the molecules responsible for its flavour are water-soluble and therefore literally go down the drain if asparagus is cooked in water.

V08 Chocolat coulant

Heston Blumenthal describes how to make a pudding containing chocolate and cheese and explains why this unlikely-sounding combination tastes good.

V09 Cooking meat

This shows experiments in which different cuts of meat are cooked under different conditions to determine the optimum cooking temperature.

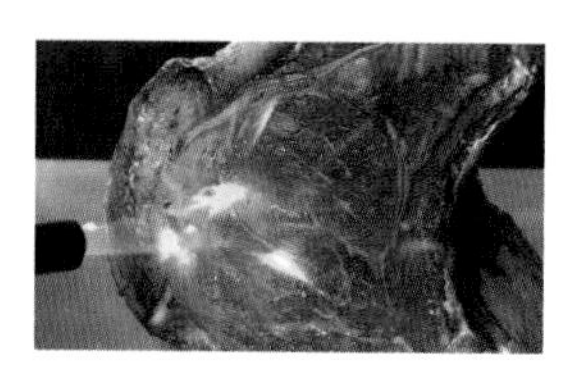

V10 Maillard browning

Browning of the surface cooked meat is caused by the Maillard reactions, which take place at temperatures well above the optimum for cooking the bulk of the meat. Heston Blumenthal uses a blowtorch to bring these reactions about.

V11 Making ice cream

Heston Blumenthal discusses the ingredients of ice cream and the pitfalls that can occur while making it – formation of lactose crystals, denaturing of egg proteins and formation of large ice crystals.

V12 The flavour of ice cream

The flavour molecules of chocolate are fat-soluble, while vanilla is water-soluble. Heston Blumenthal shows how to make chocolate and vanilla ice cream in which the chocolate flavour is released more slowly than the vanilla.

V13 The ice cream world record

Using liquid nitrogen as a coolant allows ice cream to be made in a Guinness world record time.

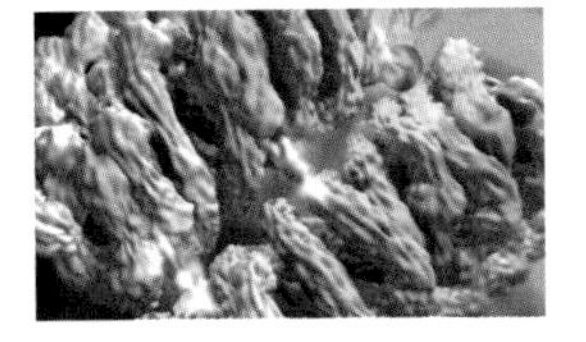

V14 The role of calcium ions

Graphics show how Ca^{2+} ions can alter the texture of foods by binding together starch and/or protein molecules.

V15 Flavour profiles

This leads on from clip V08 and shows the analysis of volatile flavour molecules by GCMS (gas chromatography mass spectrometry), leading to the idea that ingredients with several flavour components in common taste good together.

V16 Chemicals in food

Heston Blumenthal explains that all foods are made up of chemicals and shows some examples from a typical kitchen.

Kitchen**Chemistry**

The use of salt in cooking (1)

The use of salt in cooking (1)

Learning objectives

- To devise and carry out experiments to test simple hypotheses.
- To develop the idea of the 'fair test'.

Level

Age: 7–14.

Timing

One lesson of approximately one hour. Some of the planning could be done for homework.

Description

A discussion of why salt is added to water when cooking vegetables leads on to open-ended investigations devised by the students to discover (a) whether this salt can be detected in cooked beans and (b) what level of salt can be detected by taste.

Teaching notes

The lesson could be started by introducing Heston Blumenthal as a 'scientific chef' and showing the video clip of him discussing the use of salt in cooking food and/or by discussing some of the issues listed below at a level appropriate for the students concerned.

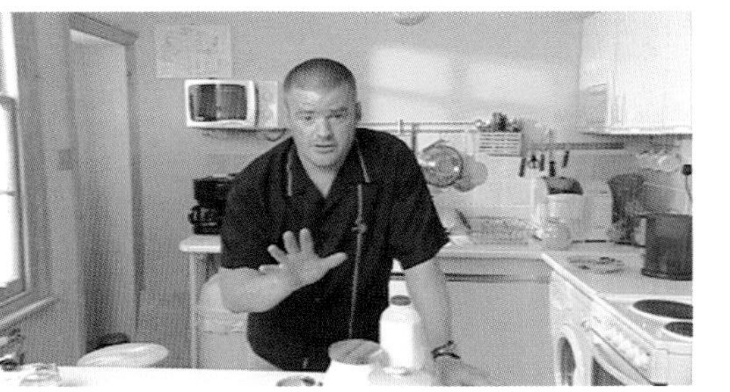

Index V01

Heston Blumenthal in the kitchen

One question posed by Heston Blumenthal early in his career as a 'scientific chef' was ***'Why do cooks add salt (sodium chloride) when cooking vegetables, for example green beans?'*** Possible reasons suggested by cooks included:

- it keeps the beans green
- it raises the boiling point of water so the beans cook faster
- it prevents the beans going soggy
- it improves the flavour.

A scientist colleague replied that there seemed to be no good reason because:

- only the acidity and calcium content of the water affect the colour of the beans
- adding salt ***does*** increase the boiling point of water but by such a small amount that it will make no difference to cooking times
- vegetables will go soggy if cooked for too long whether salt is added or not
- very little salt is actually absorbed onto the surface of a bean during cooking – typically 1/10 000 g of salt per bean which is too little to be tasted by most people.

Students can try to test some of these suggestions and explanations by experiment.

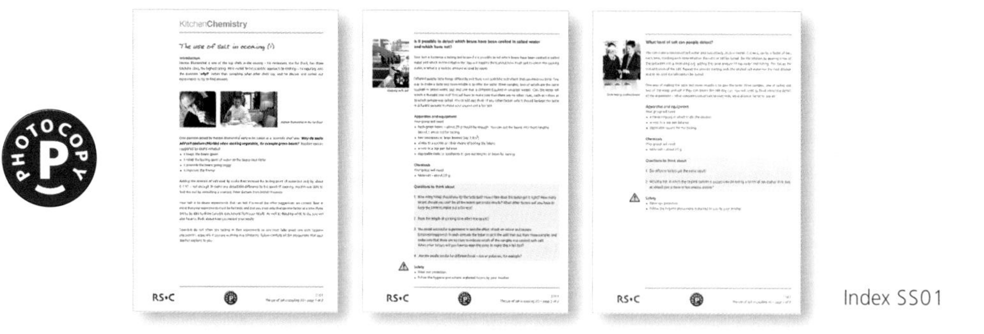

Index SS01

Is it possible to detect which beans have been cooked in salted water and which have not?

Ask students to devise a tasting test to see if it is possible to detect which beans have been cooked in salted water and which in non-salted water.

Students will need to consider how much salt to use in the cooking water, *ie* what is a realistic amount as used by cooks.

Taste testing

Taste is very subjective. One way to make a taste test more reliable is to offer the taster three samples, two of which are the same (cooked in salted water, say) and one that is different (cooked in unsalted water). Can the tester tell which is the odd one out? The experimenters will have to make sure that there are no other clues as to which sample was salted. If colour is suggested as a possible clue, blindfold tasting could be used. Other factors such as length of cooking time will also need to be controlled for a fair test.

Apparatus and equipment

Each group of students will need:

- fresh green beans – about 20 g should be enough. The beans can be cut into short lengths (about 1 cm or so) for tasting
- access to two saucepans or large beakers (say 1 dm^3)
- access to a cooker or other means of boiling the beans
- access to a top pan balance
- disposable forks or toothpicks can be used to distribute lengths of bean for tasting.

Chemicals

Each group of students will need:

- Salt – about 20 g. Buy table salt rather than 'laboratory' sodium chloride.

Questions for students

These questions are in the students' sheet and may be used to stimulate discussion.

1. How many times should you try the taste test? How often does the tester get it right? How many testers should you use? Do all the testers get similar results? What other factors will you have to keep the same to make this a fair test?

2. Does the length of cooking time affect the results?

3. You could use similar experiments to test the effect of salt on colour and texture (crispness/sogginess). In each case ask the tester to pick the odd man out from three samples and make sure that there are no clues to indicate which of the samples was cooked with salt. What other factors will you have to keep the same to make this a fair test?

4. Are the results similar for different foods – rice or potatoes, for example?

Safety

- Wear eye protection.
- Make sure that students follow proper hygiene precautions (see the section ***Experiments with food*** in ***How to use this material*** on page vii).
- Your employer's risk assessment should be consulted before carrying out this activity. This activity is covered by model (general) risk assessments widely adopted for use in UK schools and colleges such as those provided by CLEAPSS, SSERC, ASE and DfES. Bear in mind, however, that these may need some modification to suit local conditions.

What level of salt can people detect?

Students can make a solution of salt water and successively dilute it, say by a factor of two each time, checking after each dilution whether the salt can still be tasted. The dilution does not need to be done with great precision. The amount of salt usually used for cooking is typically of the order of 5 g dm^{-3} but this may vary considerably between different cooks. Students may wish to start with a more concentrated solution than this. One way of making the taste test more reliable is to give the taster three samples, one of salted and two of tap water and ask if they can detect the odd one out.

There are possibilities for much discussion about the details of the experiment – what concentration of salt to start with, what dilution factor to use *etc*.

Apparatus and equipment

Each group of students will need:

- a measuring jug in which to do the dilution. It is better to use a 'kitchen-type' jug from the Food Technology Department rather than a 'chemistry-type' measuring cylinder
- access to a top pan balance
- disposable spoons for the tasting.

Chemicals

Each group of students will need:

- salt – about 20 g. Buy table salt rather than 'laboratory' sodium chloride.

Questions for students

These questions are in the student's sheet and may be used to stimulate discussion.

1. Do different testers get the same result?

2. Would a test in which the original sample is successively diluted by a factor of ten (rather than two as above) give a more or less precise answer?

Safety

- Wear eye protection.
- Make sure that students follow proper hygiene precautions (see the section ***Experiments with food*** in ***How to use this material*** on page VII).
- Your employer's risk assessment should be consulted before carrying out this activity. This activity is covered by model (general) risk assessments widely adopted for use in UK schools such as those provided by CLEAPSS, SSERC, ASE and DfES. Bear in mind, however, that these may need some modification to suit local conditions.

Further information

Adding salt (or any other solute) does raise the boiling point of water. Boiling point elevation is a colligative property, that is it depends on the number of particles added, not their nature. Addition of 1 mole of any solute to 1 kg of water raises the boiling point by 0.52 °C. Typically a cook might add about 5 g to about 1 dm^3 of water – the equivalent of approximately 0.1 mole per kg (the molar mass of sodium chloride is 58.5 g). Bearing in mind that sodium chloride dissociates fully into sodium and chloride ions and therefore 1 mole yields 2 moles of particles, this would increase the boiling point by only 0.1 °C – not enough to affect the cooking time noticeably.

Trials suggest that salt can be tasted down to levels of roughly 0.5 g dm^{-3} but this may vary considerably between tasters. If students start with a concentration of salt of 50 g dm^{-3}, about six or seven dilutions by a factor of two will be required before the threshold of tasting is reached.

Extensions

Students could also try to devise investigations to see whether salt affects the colour or texture of cooked beans. These could be carried out if time is available or used as planning exercises.

Teacher's notes:

....

....

....

....

....

....

....

....

....

....

....

....

....

....

....

....

....

....

....

....

....

....

....

....

Kitchen**Chemistry**

The use of salt in cooking (2)

Learning objectives
- To carry out titrations to measure the concentration of Cl^- ions in aqueous solution.
- To process the results of titrations.

Level
Age: post-16.

Timing
One lesson of approximately one hour. Some of the calculations could be done for homework.

Description
A discussion of why salt is added to cooking water for vegetables leads on to a titration-based measurement of the concentration of chloride ions (and hence sodium chloride) in water before and after cooking vegetables.

Teaching notes
The lesson could be introduced by showing the video clip of Heston Blumenthal talking about the use of salt in cooking food and/or by discussing the issues listed below.

Index V01

Heston Blumenthal in the kitchen

One question posed by Heston Blumenthal early in his career as a 'scientific chef' was ***'Why do cooks add salt (sodium chloride) when cooking vegetables, for example green beans?'*** Possible reasons suggested by cooks included:
- it keeps the beans green
- it raises the boiling point of water so the beans cook faster
- it prevents the beans going soggy
- it improves the flavour.

A scientist colleague replied that there seemed to be no good reason because:
- only the acidity and calcium content of the water affect the colour of the beans
- adding salt ***does*** increase the boiling point of water but by such a small amount that it will make no difference to cooking times
- vegetables will go soggy if cooked for too long whether salt is added or not
- very little salt is actually absorbed onto the surface of a bean during cooking – typically 1/10 000 g of salt per bean which is too little to be tasted by most people.

Students can titrate the chloride ions in water with silver nitrate to investigate the hypothesis that salt is absorbed into vegetables from the cooking water during the cooking process.

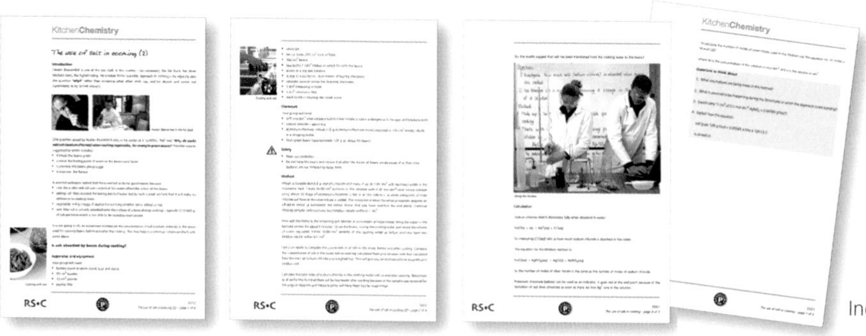

Index SS02

Is salt absorbed by beans during cooking?

Salt is sodium chloride, NaCl. On dissolving in water it dissociates to form Na^+ (sodium) and Cl^- (chloride) ions:

$$NaCl(s) \rightarrow Na^+(aq) + Cl^-(aq)$$

One way of measuring the concentration of chloride ions (and therefore the concentration of sodium chloride, since 1 mol sodium chloride produces 1 mol of chloride ions) is to titrate with silver nitrate. The equation is:

$$NaCl(aq) + AgNO_3(aq) \rightarrow AgCl(s) + NaNO_3(aq)$$

Potassium chromate solution (yellow) can be used as an indicator; it goes red at the end point because of the formation of red silver chromate as soon as there are free Ag^+ ions in the solution. In practice the end point is when the white precipitate acquires an off-white colour (a permanent red colour shows that you have overshot the end point).

Students make a solution of salt water using the amount of salt usually used for cooking (they will need to make a sensible estimate of the amount typically used) and titrate samples of this to check its concentration. They then cook some green beans in this water, strain them and record the volume of the remaining cooking water. (This is necessary because some water will have been lost by evaporation and some may have been absorbed into the beans.) They then titrate samples of the cooking water to see if there is any change in the salt concentration. Any salt lost has presumably been absorbed by the beans. Students could be given the 'recipe' for the titration given below and in the student's worksheet. Alternatively, they could consider what concentration of silver nitrate to use to give reasonable titration figures.

Apparatus and equipment

Each group of students will need:

- burette stand or retort stand, boss and clamp
- 50 cm^3 burette
- 10 cm^3 pipette
- pipette filler
- white tile
- two or three 250 cm^3 conical flasks

- 100 cm^3 beaker
- saucepan or 1 dm^3 beaker in which to cook the beans
- access to a top pan balance
- access to a cooker or other means of boiling the beans
- colander, sieve or similar for straining the beans
- 1 dm^3 measuring cylinder
- 1 dm^3 volumetric flask
- wash bottle containing deionised water.

The teacher may wish to have a bottle for collection of silver residues.

Chemicals

Each group of students will need:

- 0.05 mol dm^{-3} silver nitrate solution (silver nitrate solution is dangerous to the eyes and blackens skin)
- sodium chloride – about 6 g
- potassium chromate indicator (5 g potassium chromate (toxic) dissolved in 100 cm^3 water), ideally in a dropping bottle
- fresh green beans (approximately 125 g or about 50 beans).

Safety

- Wear eye protection.
- Warn students not to taste the beans and ensure that after the lesson all beans are disposed of so that other students are not tempted to taste them.
- Your employer's risk assessment should be consulted before carrying out this activity. This activity is covered by model (general) risk assessments widely adopted for use in UK schools and colleges such as those provided by CLEAPSS, SSERC, ASE and DfES. Bear in mind, however, that these may need some modification to suit local conditions.

Method

Weigh accurately about 6 g sodium chloride and make it up to 1.00 dm^3 with deionised water in the volumetric flask. Titrate 10.00 cm^3 portions of this solution with 0.05 mol dm^{-3} silver nitrate solution using about 10 drops of potassium chromate solution as the indicator. A white precipitate of silver chloride will form as the silver nitrate is added. The end point is when the white precipitate acquires an off-white colour (a permanent red colour shows that you have overshot the end point). Continue titrating samples until you have two titration results within 0.1 cm^3.

Now add the beans to the remaining salt solution in a saucepan or large beaker. Bring the water to the boil and simmer for about 5 minutes. Strain the beans, saving the cooking water, and record the volume of water recovered. Titrate 10.00 cm^3 samples of the cooking water as before until you have two titration results within 0.1 cm^3.

Use your results to calculate the concentration of salt in the water before and after cooking. Compare the concentration of salt in the water before cooking calculated from your titration with that calculated from the mass of sodium chloride you weighed out. This will give you an estimate of how accurate your titration was.

Calculate the total mass of sodium chloride in the cooking water before and after cooking. Remember to allow for the fact that there will be less water after cooking because of the samples you removed for the original titrations and because some will have been lost by evaporation.

Do the results suggest that salt has been transferred from the cooking water to the beans?

Doing the titration

Typical results

Cooking green beans

6.54 g table salt (sodium chloride) was made up to 1.00 dm^3 in deionised water (this is the equivalent of a 0.118 mol dm^{-3} solution). 10.00 cm^3 portions were titrated with 0.05 mol dm^{-3} silver nitrate solution with a few drops of potassium chromate indicator. Titres of 21.20 and 21.12 cm^3 were obtained (average value 21.16 cm^3). This gives the concentration of Cl^- in the solution as 0.1058 mol dm^3.

131.89 g of fine beans (50 beans) was placed in the salt solution (980 cm^3 after the two titration samples were allowed for) and the water was brought to the boil (15 minutes) and boiled for a further 5 minutes. The beans were then drained and the cooking water recovered. 980 cm^3 was recovered. (The beans did not taste noticeably salty.)

10.00 cm^3 portions of the recovered cooking water were titrated with 0.05 mol dm^{-3} silver nitrate solution with a few drops of potassium chromate indicator. Titres of 22.51, 22.69 and 22.75 cm^3 were obtained. Average of the last two titres = 22.72 cm^3.

Calculation

moles Cl^- in 10.00 cm^3 portions of the recovered cooking water = (0.05 x 22.72)/1000 = 1.136×10^{-3} mol

moles Cl^- in 980 cm^3 recovered cooking water = 1.136×10^{-3} x 98 = 0.1113 mol

grams NaCl in 980 cm^3 recovered cooking water = 0.1113 x 58.5 = 6.51 g

grams NaCl in 980 cm^3 water before cooking = 0.1058 x 58.5 x 0.98 = 6.07 g (based on titration)

grams NaCl in 980 cm^3 water before cooking = 6.54 x 0.98 = 6.41 g (as weighed out)

Cooking potatoes

6.34 g table salt (NaCl) was dissolved in 1.00 dm^3 distilled water (this is the equivalent of a 0.1083 mol dm^{-3} solution). 10.00 cm^3 portions were titrated with 0.05 mol dm^{-3} $AgNO_3$ with a few drops of potassium chromate indicator. Titres of 21.26 and 21.32 cm^3 (average 21.29 cm^3) were obtained. This gives the concentration of Cl^- in the solution as 0.1065 mol dm ^{3}.

230.24 g of new potatoes (20 pieces with skin left on but cut to approx. the size of a marble) were placed in the salt solution (980 cm^3 after the two titration samples were allowed for) and the water was brought to the boil (14 minutes) and boiled for a further 10 minutes. The potatoes were then drained and the cooking water recovered. 910 cm^3 was recovered. (The potatoes tasted slightly salty.)

10.00 cm^3 portions of the cooking water were titrated with 0.05 mol dm^{-3} $AgNO_3$ with a few drops of potassium chromate indicator. Titres of 22.85 and 23.20 cm^3 were obtained. (Average 23.025 cm^3)

Calculation

moles Cl^- in 10.00 cm^3 portions of the recovered cooking water = (0.05 x 23.025)/1000 = 1.151×10^{-3} mol

moles Cl^- in 910 cm^3 recovered cooking water = 1.151×10^{-3} x 91 = 0.1047 mol

grams NaCl in 910 cm^3 recovered cooking water = 0.1047 x 58.5 = 6.13 g

grams NaCl in 980 cm^3 water before cooking = 0.1065 x 58.5 x 0.98 = 6.11 g (based on titration)

grams NaCl in 980 cm^3 water before cooking = 6.21 g (as weighed out)

The results for both beans and potatoes suggest that if anything there is a slight increase in the amount of salt in the water after cooking compared with before. Certainly they do not support the hypothesis that vegetables absorb significant amounts of salt from the cooking water. If anything, it appears that salt might have been leached out of the vegetable during cooking rather than being absorbed from the cooking water. This could be a point for discussion.

Further information

The amount of salt typically used to cook vegetables is of the order of 5 g dm^{-3}, although this may vary considerably from cook to cook. This is very approximately 0.1 mol dm^{-3} (M_r NaCl = 58.5). If a 10 cm^3 sample of cooking water is used, and titrated with 0.05 mol dm^{-3} silver nitrate solution, this would give a titre of approximately 20 cm^3.

Most table salt contains small amounts of additives such as magnesium carbonate and potassium hexacyanoferrate(II) which prevent caking. This could affect the titration results and would help to account for the discrepancy between the weighed amount of sodium chloride and the amount measured by titration.

Students could compare the amount of salt in the cooking water determined by titration with that weighed out to give an estimate of experimental error to help them discuss the significance of their results.

This activity is most suitable for post-16 students but it could be used with Key Stage 4 (or equivalent) students. Such students might find the calculations daunting. One possibility to simplify the calculation would be for the teacher to set up a spreadsheet with the conversion factor from titre to mg salt per dm^3 cooking water already entered so that students simply have to enter their titre. The formula is: titre x 0.293.

Extension

Beans could be cooked for different lengths of time. What is the effect of increasing the cooking time on the level of salt remaining in the cooking water?

Students could try different types of food, *eg* peas, rice, potatoes, pasta – is there a significant difference in the level of salt remaining in the cooking water?

This could be an opportunity for discussion about controlling variables and about fair tests.

The level of salt in the vegetable before and after cooking could be measured directly by finely chopping or pureeing a weighed quantity of the vegetable and boiling with deionised water to extract the salt into the water and then carrying out a titration.

This titration method can be adapted to find the salt content of different foods. One source, *The Most Useful Science* Harleston: Thorburn Kirkpatrick, 1989, quotes the following method.

1. Place 25 g of minced food sample in a 250 cm^3 flask.

2. Add water to make up to a 10% aqueous dispersion.

3. Mix thoroughly and allow to stand for 30 minutes. Remix and filter.

4. Pipette 25 cm^3 of filtrate into a conical flask and add 0.5 cm^3 of potassium chromate solution.

5. Titrate against 0.1 mol dm^{-3} silver nitrate solution.
   $NaCl + AgNO_3 \rightarrow AgCl + NaNO_3$

   1 cm^3 of 0.1 mol dm^{-3} $AgNO_3$ is equivalent to 0.00585 g NaCl.

   Assuming 25 cm^3 of filtrate is equivalent to 2.5 g food (for 10% dispersion) then salt g per 100 g food = 0.00585 x titre x 100 / 2.5

Questions for students

1. What assumptions are being made in this method?

2. What is assumed to be happening during the 30 minutes in which the dispersion is left standing?

3. Explain why '1 cm^3 of 0.1 mol dm^{-3} $AgNO_3$ is equivalent to 0.00585 g NaCl'.

4. Explain how the equation
   salt g per 100 g food = titre x 0.00585 x 40
   is arrived at.

Answers

1. It is assumed that all the salt in the food will dissolve in the added water.

2. Salt is dissolving in the water.

3. From the equation for the titration, 1 mol $AgNO_3$ reacts with 1 mol (58.5) of NaCl
   1 cm^3 of 0.1 mol dm^{-3} $AgNO_3$ contains 1 x 0.1 / 1000 mol = 1×10^{-4} mol $AgNO_3$
   This is equivalent to 58.5×10^{-4} = 0.000585 g NaCl

4. titre x 0.00585 is the number of grams of salt in 25 cm^3 of dispersion

   titre x 0.00585 x 10 is the number of grams of salt in 250 cm^3 of dispersion (25 g of food)

   titre x 0.00585 x 10 / 25 is the number of grams of salt in 1 g of food titre x 0.00585 x 10 x 100 / 25 is the number of grams of salt in 100 g of food

titre x 0.00585 x 40 is the number of grams of salt in 100 g of food

Teacher's notes:

....

Kitchen**Chemistry**

By how much does salt increase the boiling point of water?

By how much does salt increase the boiling point of water?

Learning objectives

- To devise and carry out experiments to test simple hypotheses.
- To develop the idea of the 'fair test'.

Level

Age: 11–14.

Timing

One lesson of approximately one hour. Graphs could be done for homework.

Description

Students devise and carry out an experiment to measure the increase in boiling point brought about by adding sodium chloride to water, thus helping to test one of the suggestions as to why cooks add salt to the water when cooking vegetables.

Teaching notes

The lesson could be started by introducing Heston Blumenthal as a 'scientific chef' and showing the video clip of him discussing the use of salt in cooking food and/or by discussing some of the issues listed below at a level appropriate for the students concerned.

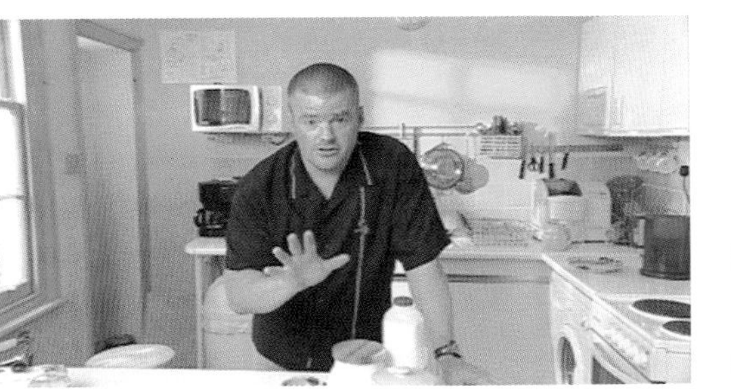

Index V01

Heston Blumenthal in the kitchen

One question posed by Heston Blumenthal early in his career as a 'scientific chef' was ***'Why do cooks add salt (sodium chloride) when cooking vegetables, for example green beans?'*** Possible reasons suggested by cooks included:

- it keeps the beans green
- it raises the boiling point of water so the beans cook faster
- it prevents the beans going soggy
- it improves the flavour.

A scientist colleague replied that there seemed to be no good reason because:

- only the acidity and calcium content of the water affect the colour of the beans
- adding salt ***does*** increase the boiling point of water but by such a small amount that it will make no difference to cooking times
- vegetables will go soggy if cooked for too long whether salt is added or not
- very little salt is actually absorbed onto the surface of a bean during cooking – typically 1/10 000 g of salt per bean which is too little to be tasted by most people.

Students might be able to find how adding salt affects the boiling point of water in a reference book or on the internet. Here they do an experiment to measure the boiling point of pure water and of a series of solutions containing different amounts of salt. They draw a graph to help interpret their results. They also need to estimate the concentration of salt usually used by cooks when cooking vegetables.

An important teaching point is that a single measurement with a realistic amount of salt as used by cooks will produce only a small change in boiling point that is very difficult to measure. So, more concentrated solutions are used to produce a graph from which a result may be extrapolated.

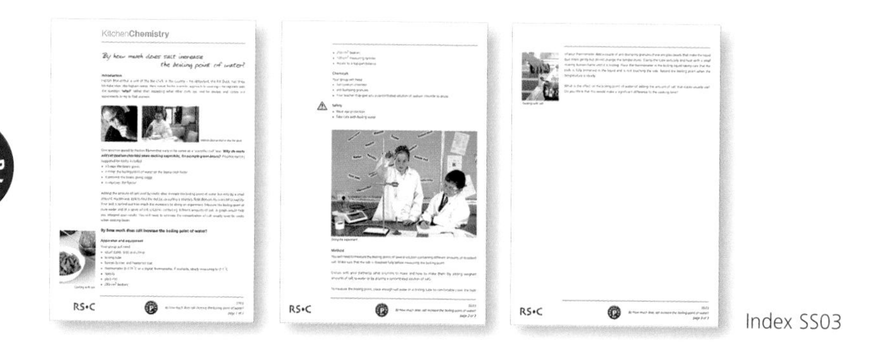

Index SS03

By how much does salt increase the boiling point of water?

Measuring the boiling point of a salt solution

Method

If simply set the task of finding the effect of added salt on the boiling point of water, most students will probably just make solutions with different amounts of salt and measure the boiling points. This has three disadvantages:

- it will use a lot of salt
- the salt takes some time to dissolve fully
- unless quite concentrated solutions are used, it will be difficult to measure any difference in the boiling point.

By how much does salt increase the boiling point of water?

Teachers may wish students to find this out for themselves but a better alternative may be to start with a concentrated (almost saturated) solution of salt and successively dilute this. If this method is to be adopted, the dilution procedure may need to be explained to the students.

Boiling points are most conveniently measured by clamping a boiling tube vertically, adding a few centimetres depth of salt water (enough to comfortably cover the bulb of a thermometer) and boiling gently over a small Bunsen flame. Students must take care to ensure that the thermometer bulb is fully immersed and not touching the side of the boiling tube. A sensitive thermometer is needed and a digital thermometer (if available) may be a better alternative to a traditional mercury-in-glass one.

Alternatively, electronic thermometers with data loggers could be used if the school or college has suitable equipment.

The addition of a few anti-bumping granules will make the solutions boil more gently. Their use will have to be explained to students.

Students should plot suitable graphs (either by hand or using a spreadsheet package) and use these to estimate the increase in boiling point if a typical amount of salt as added by a cook were used.

Apparatus and equipment

Each group of students will need:

- retort stand, boss and clamp
- boiling tube
- Bunsen burner and heatproof mat
- thermometer 0–110 °C or a digital thermometer, if available, ideally measuring to 0.1 °C
- spatula
- glass rod
- 250 cm^3 beakers
- 100 cm^3 measuring cylinder
- access to a top pan balance.

Chemicals

Each group of students will need:

- salt (sodium chloride); it may be better to use cooking salt from a shop rather than sodium chloride from a chemical supplier
- anti-bumping granules.

Safety

- Wear eye protection.
- Take care with boiling water.
- Your employer's risk assessment should be consulted before carrying out this activity. This activity is covered by model (general) risk assessments widely adopted for use in UK schools and colleges such as those provided by CLEAPSS, SSERC, ASE and DfES. Bear in mind, however, that these may need some modification to suit local conditions.

Some teachers may prefer to give students a concentrated solution of sodium chloride to dilute. The solubility of sodium chloride at room temperature is approximately 6 mol dm^{-3} (approximately 360 g dm^{-3}). If this method is used, students will need to be told the concentration of the solution.

It is probably better to work in grams of salt rather than moles for this activity.

Further information

Boiling point elevation is a colligative property; that is it depends on the number of particles dissolved rather than their nature. The boiling point constant for water is 0.52 °C when 1 mole of solute is dissolved in 1 kg of water. Since sodium chloride dissociates fully into Na^+ and Cl^- ions on dissolving in water, the boiling point of water increases by 1.04 °C for every mole (58.5 g) of sodium chloride dissolved in 1 dm^3 (approximately 1 kg) of water. Typically a cook might add 5 g (*ie* approximately 1/10 mole) or so of salt to 1 dm^3 of water so this would increase the boiling point by about 0.1 °C. A saturated solution of sodium chloride is about 6 mol dm^{-3} and so would have a boiling point of about 106 °C.

The reason that added solute increases the boiling point may be explained simply as follows. Boiling involves the escape of water molecules from the surface of the liquid. To escape, a water molecule must be moving fast enough (have enough energy) and be moving towards the surface. The presence of particles of solute will hinder the escape of such molecules and make it more difficult for the water to boil.

Extension

Students could measure the boiling point elevation produced by other solutes.

Teacher's notes:

Kitchen**Chemistry**

Is all salt the same?

Learning objective

- To critically analyse advertising claims in a scientific context.

Level

Ages: 11–16.

Timing

Between 30 minutes and one hour.

Description

Students consider, critically analyse and debate statements about different types of salt used in cooking and claims made for them.

Resources

Collect cuttings or photocopies of packages of different types of salts and recipes that refer to different types of salt, and that make particular claims. The more topical these are, the better. Students could be asked to collect these and bring them to school or college. Alternatively a search of the internet should reveal many interesting items. The PowerPoint® presentation offers a selection.

Power Point

Index PP01

A selection of types of salt

Kitchen Chemistry

Is all salt the same?

Questions

RS•C

RS•C

Index SS04

Is all salt the same?

Teaching notes

Some recipes suggest the use of sea salt rather than 'ordinary' table salt. Students are asked to read the statements on the packaging of different brands of sea salt and consider to what extent they are scientifically accurate and whether they might be misleading. Some statements are given below but the teacher and/or students could collect statements from packs found at home in the kitchen, in shops, in magazines and in recipes to use in addition to these and to those on the PowerPoint® presentation. Handing out actual cuttings or photocopies would make the exercise more realistic. Other questions could be added to or substituted for those below to fit the actual statements used. Students could be asked to work individually or in groups. They could be asked to write written responses or to be prepared to contribute to a debate or discussion. The following passage is also reproduced in the student's worksheet.

One brand of sea salt (sold in the form of flakes for grinding in a mill) has the following wording on the packet:

'Ingredient Pure Crystal Sea Salt with no additives.'

'Its combination of texture and flavour sets it apart from other salts and it is sought after by the health conscious and gourmet alike.'

'XXXX is a completely natural product without artificial additives, retaining valuable sea water trace elements such as magnesium and calcium.'

'Its pronounced and distinctive 'salty' taste means less is required, an advantage for those who wish to reduce their salt intake.'

'It is free from the bitter after-taste often associated with other salts and salt substitutes.'

Students could be asked to consider the following questions

1. To what extent are the statements about the sea salt contradictory? Can it be 'pure' and at the same time 'retain valuable sea water trace elements such as magnesium and calcium'?

2. Do the statements about sea salt suggest that it has a more salty taste than ordinary salt? Is this possible or likely considering that both types of salt are almost 100% sodium chloride?

3. Most sea salts are sold in the form of granules or flakes which are larger than the granules of ordinary table salt. Could this affect the taste:
   (a) when placed directly on the tongue?
   (b) when the salt is dissolved in water as is done when cooking vegetables?

4. Sea salt is made by evaporating the water from seawater (in hot countries this if often done in outdoor lagoons using the Sun's rays). Most 'ordinary' salt in the UK is extracted from underground deposits by pumping water into the deposits to dissolve some of the salt and evaporating the resulting brine:
   (a) Which is likely to be purer?
   (b) Consider where the underground deposits came from. Can both types of salt be considered to be sea salt?

Further information

Many sea salts are coloured. This is caused by small amounts of algae and/or clays from the evaporation pond.

Students could be asked to devise (and try out) taste tests to see if sea salt does taste 'saltier' than table salt both directly on the tongue and dissolved in water. If tasting is to be attempted, appropriate hygiene and safety measures must be used, (see the section ***Experiments with food*** in ***How to use this material*** on page vii). It seems unlikely that, once dissolved in water, the ions could 'remember' which crystalline form they had once been in, and controlled tasting tests seem to confirm this. Directly on the tongue, the two types of crystal will probably dissolve at different rates and this could lead to a sensation of greater or lesser saltiness.

Salt in the form of flakes or large granules will not pack together as well as the small granules of table salt so, if salt is added to a recipe by volume (*eg* a spoonful) rather than by weight, less salt will be used and a claim that 'less is required' could perhaps be justified. (Although once this is realised, it would be cheaper to simply use less ordinary table salt.) This is the explanation behind claims such as *'YYYY is pure salt (sodium chloride) but contains 33% less sodium per teaspoon'*. This brand of salt contains just as much sodium (39.3% by weight) as any other sodium chloride – it is simply that a teaspoonful of its flaky crystals contains 33% less sodium chloride than a teaspoonful of conventional granulated table salt.

Taste testing is inevitably subjective, but one way in which it can be made more reliable is for the taster to be offered three samples, two of which are the same, and see if they can pick the odd one out.

Answers

Because of the nature of the question, there are no right and wrong answers.
The following are some suggestions.

1. This question is about the everyday usage of the word 'pure', which is often used to mean something like 'wholesome' and its scientific usage which means a single substance unadulterated by anything else. Salt containing trace elements as well as sodium chloride cannot be pure in the scientific sense.

2. Some statements made about some brands of salt do seem to suggest that sea salt has a more salty taste than ordinary salt. It seems unlikely that one sample of pure sodium chloride could be 'saltier' than another unless this is caused by different sizes of crystals dissolving at different rates, see question 3.

3. Different sized crystals might dissolve on the tongue at different rates and thus give a more or less salty taste. No such effect seems possible when salt is dissolved in water.

4. (a) Seawater evaporating in lagoons is likely to suffer contamination from a whole range of sources. That made by dissolving underground deposits is likely to be purer as the deposited salt has in effect been recrystallised when it formed and on re-dissolving, any insoluble impurities will be discarded.

   (b) The underground deposits were made by the drying up of ancient seas, so both types can be considered to be sea salt.

Teacher's notes:

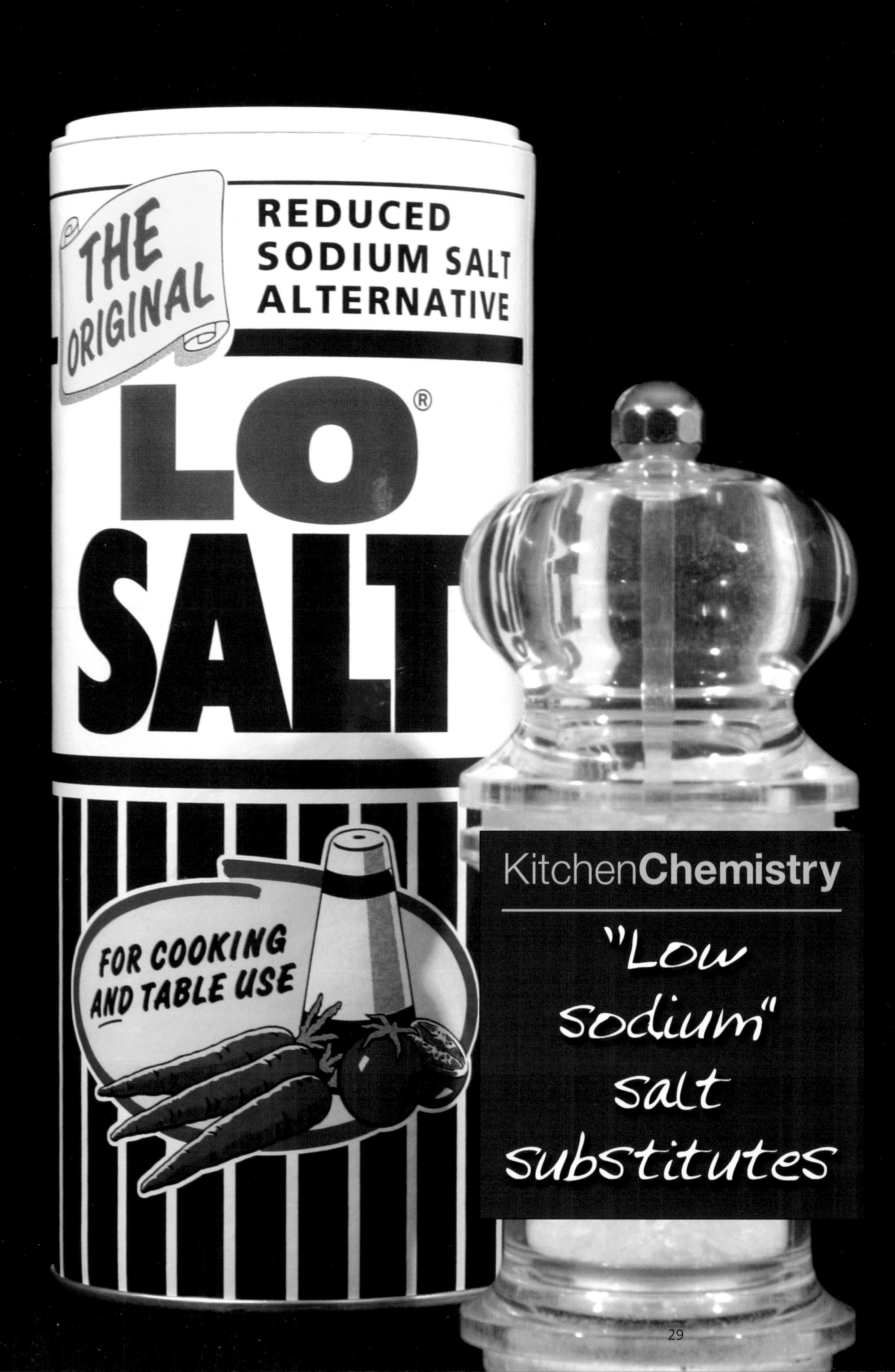
THE ORIGINAL
REDUCED SODIUM SALT ALTERNATIVE
LO SALT®
FOR COOKING AND TABLE USE
KitchenChemistry
"Low sodium" salt substitutes

"Low Sodium" salt substitutes

Learning objectives

- To carry out titrations to measure the concentration of Cl^- ions in aqueous solution.
- To process the results of titrations.

Level

Age: post-16.

Timing

One lesson of approximately one hour.

Description

Students use a titration procedure to measure the proportions of potassium chloride and sodium chloride in 'low salt' salt substitutes.

Teaching notes

Salt (sodium chloride) is a vital component of our diet both for our health and for its flavouring effect. In appropriate quantities it is needed for transmission of nerve impulses and for contraction of muscles, although in excessive amounts it is associated with heart disease, high blood pressure and strokes.

Some people, who for health reasons (high blood pressure, for example) require a diet low in sodium, use salt substitutes – one trade name is LoSalt. Further information is available at *www.losalt.com* (*accessed January 2005*), for example. These products are either potassium chloride (KCl) or mixtures of sodium chloride (NaCl) and potassium chloride. The titration method described enables students to measure the percentage of potassium chloride in these mixtures. A suitable method is described in the student's worksheet and is reproduced below.

KitchenChemistry

"Low Sodium" salt substitutes

RS•C

RS•C

RS•C

Index SS05

On dissolving in water both potassium chloride and sodium chloride dissociate completely to form Cl^- (chloride) ions, *eg*:

$NaCl(s) \rightarrow Na^+(aq)$ and $Cl^-(aq)$

One way of measuring the total concentration of chloride ions is to titrate with silver nitrate. The equation is:

$Cl^-(aq) + AgNO_3(aq) \rightarrow AgCl(s) + NO_3^-(aq)$

Potassium chromate solution (yellow) can be used as an indicator; it goes red at the end point because of the formation of red silver chromate as soon as there are free Ag^+ ions in the solution. In practice the end point is when the white precipitate acquires an off-white colour (a permanent red colour shows that you have overshot the end point).

The calculation is somewhat unusual. It depends on the fact that a given mass of potassium chloride will contain less chloride than the same mass of sodium chloride because the potassium atom is more massive than the sodium atom. The calculation is illustrated by the following example.

0.10 g of a salt substitute (a mixture of sodium chloride and potassium chloride) was dissolved in water and titrated with 0.05 mol dm^{-3} silver nitrate solution using potassium chromate as indicator. 30.52 cm^3 of silver nitrate solution was required.

Using the relationship moles of solute = M x v/1000, where M = concentration in mol dm^{-3} and v = volume in cm^3

moles $AgNO_3$ = moles Cl^- = 0.05 x 30.52/1000 = 1.526×10^{-3}

If the mixture had been 100% NaCl (M_r = 58.5):

moles Cl^- = 0.10/58.5 = 1.709×10^{-3}

If the mixture had been 100% KCl (M_r = 74.5):

moles Cl^- = 0.10/74.5 = 1.342×10^{-3}

Let % KCl be x, then % NaCl = 100-x

$$\text{Moles } Cl^- \text{ in mixture} = \frac{(1.342 \times 10^{-3})\,x + 1.709 \times 10^{-3}\,(100-x)}{100}$$

$$1.526 \times 10^{-3} = \frac{(1.342 \times 10^{-3})\,x + 1.709 \times 10^{-3}\,(100-x)}{100}$$

$$152.6 = 1.342\,x + 170.9 - 1.709\,x$$

$$18.3 = 0.367\,x$$

$$x = 50$$

So the mixture was 50% potassium chloride and 50% sodium chloride.

Alternatively, the calculation could be set out as follows.

100% KCl 1.342×10^{-3} mol   Actual 1.526×10^{-3} mol   100% NaCl 1.709×10^{-3} mol

|← 0.184×10^{-3} mol →|← 0.183×10^{-3} mol →|

|← 0.367×10^{-3} mol →|

Our measured no. of moles of Cl^- is $(0.183 \times 10^{-3}/0.367 \times 10^{-3}) \times 100\% = 50\%$ of the way from pure NaCl to pure KCl, so the mixture was 50% NaCl, 50% KCl.

Apparatus and equipment

Each group of students will need:

- burette stand or retort stand, boss and clamp
- 50 cm^3 burette
- 10 cm^3 pipette
- pipette filler
- white tile
- two or three 250 cm^3 conical flasks
- 100 cm^3 beaker
- access to a top pan balance
- 100 cm^3 volumetric flask
- wash bottle containing deionised water.

Chemicals

Each group of students will need:

- 0.05 mol dm^{-3} silver nitrate solution (silver nitrate solution is dangerous to the eyes and blackens skin) – about 100 cm^3
- a sample of salt substitute such as LoSalt, preferably in a branded container, or a mixture of sodium chloride and potassium chloride of known composition
- potassium chromate indicator (5 g potassium chromate (toxic) dissolved in 100 cm^3 water), ideally in a dropping bottle.

Safety

- Wear eye protection.
- Your employer's risk assessment should be consulted before carrying out this activity. This activity is covered by model (general) risk assessments widely adopted for use in UK schools and colleges such as those provided by CLEAPSS, SSERC, ASE and DfES. Bear in mind, however, that these may need some modification to suit local conditions.

Method

Weigh out accurately about 1 g of salt substitute containing a mixture of sodium chloride and potassium chloride and make it up to 100.00 cm^3 with deionised water in a volumetric flask. Titrate 10.00 cm^3 portions of this solution with 0.05 mol dm^{-3} silver nitrate solution using about 10 drops of potassium chromate solution as the indicator. A white precipitate of silver chloride will form as the silver nitrate is added. The end point is when the white precipitate acquires an off-white colour (a permanent red colour shows that you have overshot the end point). Continue titrating samples until you have two titration results within 0.1 cm^3.

Typical results

The Lo Salt pack states 66% (min) KCl, 33.3% NaCl, $MgCO_3$ anti-caking agent.

1.00 g of Lo Salt was made up to 100 cm^3 in deionised water. 10.00 cm^3 portions were titrated with 0.05 mol dm^{-3} $AgNO_3$ with a few drops of potassium chromate indicator. Titres of 29.72, 29.60 and 29.60 cm^3 were obtained.

Calculation

Moles $AgNO_3$ = moles Cl^- = (M x V)/1000 = (29.60 x 0.05)/1000 = 1.48 x 10^{-3}

If the mixture had been 100% NaCl (M_r = 58.5):

moles Cl^- = 0.10/58.5 = 1.709 x 10^{-3}

If the mixture had been 100% KCl (M_r = 74.5):

moles Cl^- = 0.10/74.5 = 1.342 x 10^{-3}

Let % KCl be x, then % NaCl = 100-x

$$\text{Moles } Cl^- \text{ in mixture} = \frac{(1.342 \times 10^{-3})\,x + 1.79.0 \times 10^{-3}\,(100-x)}{100}$$

$$1.480 \times 10^{-3} = \frac{(1.342 \times 10^{-3})\,x + 1.79.0 \times 10^{-3}\,(100-x)}{100}$$

$$148.0 = 1.342\,x + 170.9 - 1.709\,x$$

$$22.9 = 0.367\,x$$

$$x = 62.4$$

So the mixture was 62.4% potassium chloride and 37.6% sodium chloride, close to the percentages stated on the pack.

Alternatively, the calculation could be set out as below:

100% KCl 1.342 x 10^{-3} mol   Actual 1.480 x 10^{-3} mol   100% NaCl 1.709 x 10^{-3} mol

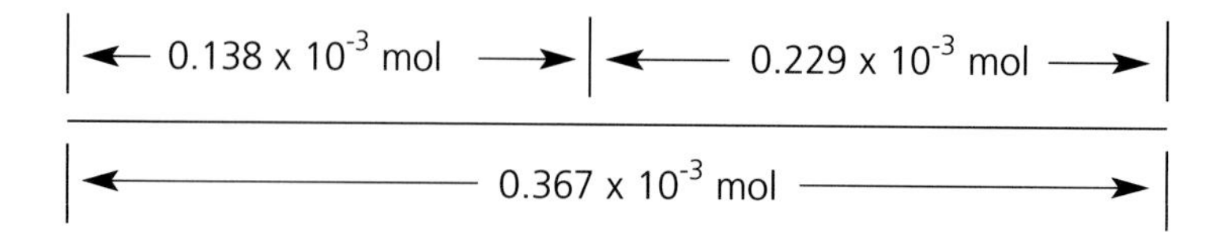

The measured no. of moles of Cl^- is ($0.138 \times 10^{-3}/0.367 \times 10^{-3}$) x100% = 37.6% of the way from pure NaCl to pure KCl so the mixture was 37.6% NaCl, 62.4% KCl, close to the percentages stated on the pack.

Further information

Some salt substitutes contain small amounts of anti-caking agents such as magnesium carbonate which might affect the results.

Extension

Some recipes specify baking fish, for example, in rock salt (an impure form of sodium chloride containing insoluble impurities, mostly sand). Students could discuss whether this is likely to be more or less effective than pure salt and could measure the percentage of sodium chloride either by the silver nitrate titration method (post-16 level) or by the traditional filtration method for separating salt from sand (pre-16). In the latter method the amount of salt can be found by difference.

Teacher's notes:

....

....

....

....

....

....

....

....

....

....

....

....

....

....

....

....

Kitchen**Chemistry**

What affects the colour and texture of cooked vegetables?

RS•C

What affects the colour and texture of cooked vegetables?

Learning objectives

- To devise and carry out experiments to test simple hypotheses.
- To develop the idea of control experiments.
- To revise some simple organic and inorganic reactions.
- To revise some aspects of hydrogen bonding.

Level

Age: post-16.

Timing

At least one lesson of approximately one hour for each section attempted.

Description

The lesson could be started by introducing Heston Blumenthal as a 'scientific chef' and showing the video clip of him discussing the use of salt in cooking food and/or by discussing some of the issues listed below at a level appropriate for the pupils concerned.

Index V01

Heston Blumenthal in the kitchen

One question posed by Heston Blumenthal early in his career as a 'scientific chef' was ***'Why do cooks add salt (sodium chloride) when cooking vegetables, for example green beans?'*** Possible reasons suggested by cooks included:

- it keeps the beans green
- it raises the boiling point of water so the beans cook faster
- it prevents the beans going soggy
- it improves the flavour.

A scientist colleague replied that there seemed to be no good reason because:

- only the acidity and calcium content of the water affect the colour of the beans
- adding salt ***does*** increase the boiling point of water but by such a small amount that it will make no difference to cooking times
- vegetables will go soggy if cooked for too long whether salt is added or not
- very little salt is actually absorbed onto the surface of a bean during cooking – typically 1/10 000 g of salt per bean which is too little to be tasted by most people.

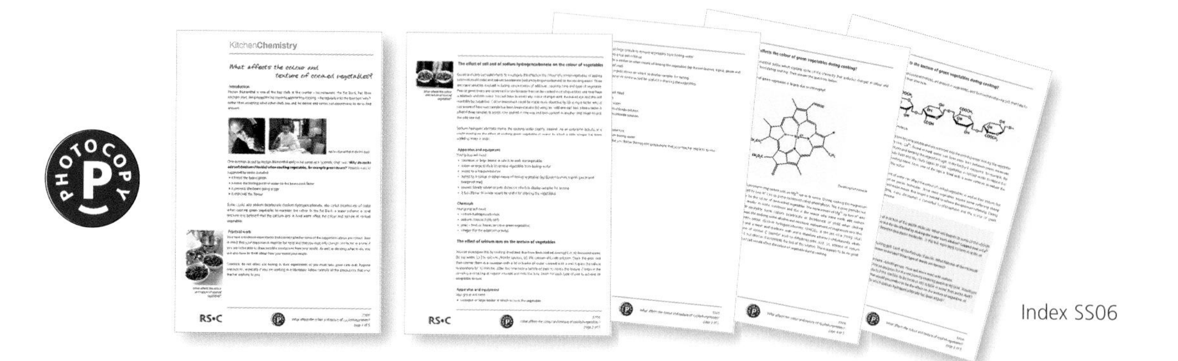

Index SS06

1. What affects the colour of green vegetables during cooking?

The colour of green vegetables is largely due to chlorophyll.

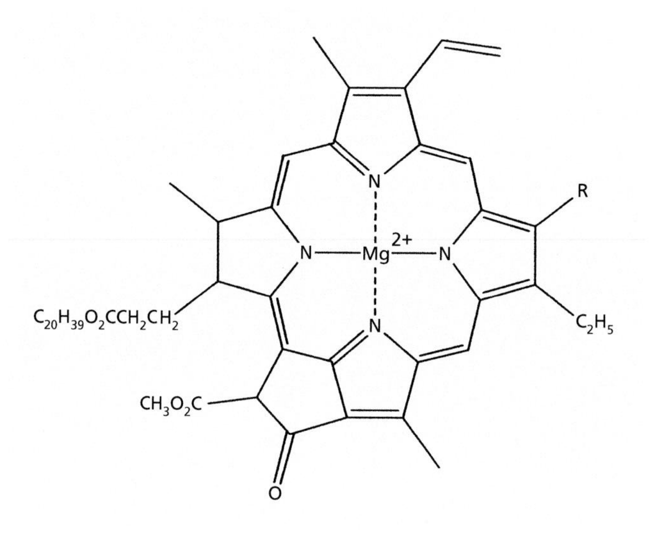

The chlorophyll molecule

Chlorophyll has a porphyrin ring system with an Mg^{2+} ion in its centre. During cooking this magnesium ion can be replaced by two H^+ ions to give a compound called phenophytin. This is olive green/brown and is responsible for the colour of overcooked vegetables. The replacement of Mg^{2+} by two H^+ ions takes place most readily in acidic conditions and this is the reason why some cooks add sodium hydrogencarbonate (everyday name sodium bicarbonate or bicarbonate of soda) when cooking vegetables. This keeps the cooking water alkaline and minimises replacement of magnesium ions thus maintaining the green colour. (Sodium hydrogencarbonate, $NaHCO_3$, is the salt of a strong alkali (sodium hydroxide) and a weak acid (carbonic acid) and is therefore alkaline in solution because of its interaction with water.) Unfortunately, alkalis catalyse the oxidation of vitamin C (ascorbic acid) to dehydroascorbic acid. So, addition of sodium hydrogencarbonate is not ideal as it accelerates the loss of this vitamin. There appears to be no good reason why addition of salt would affect the colour of vegetables during cooking.

2. What affects the texture of green vegetables during cooking?

Pectins, which are polysaccharides, are present in vegetables, and form water-retaining gels that help to give vegetables their structure.

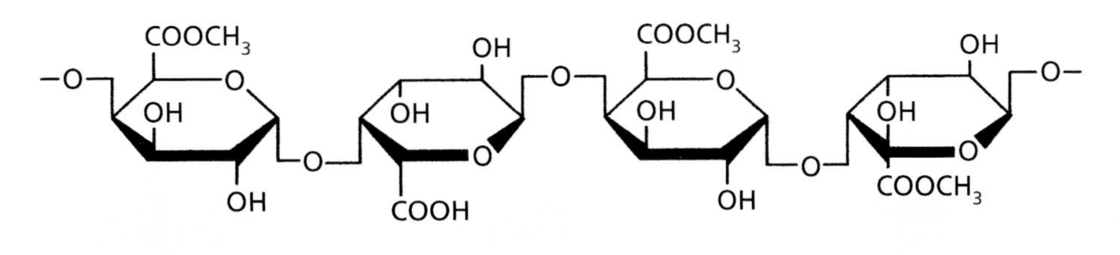

A section of a pectin molecule

During cooking, pectins become soluble and are extracted into the cooking water making the vegetable go mushy. Calcium ions, Ca^{2+}, found in hard water, can form cross links between pectin molecules making them less soluble and keeping the vegetable tough. In the Fat Duck restaurant, for example, the local water is relatively hard and the chefs began to cook vegetables in bottled water to reduce this effect and shorten cooking times. Now, one of the taps is fitted with a water softener to reduce the level of Ca^{2+} ions in the water. There is a brief mention of the part played by calcium ions in binding both carbohydrate and protein molecules in the following video clip.

Video Clip

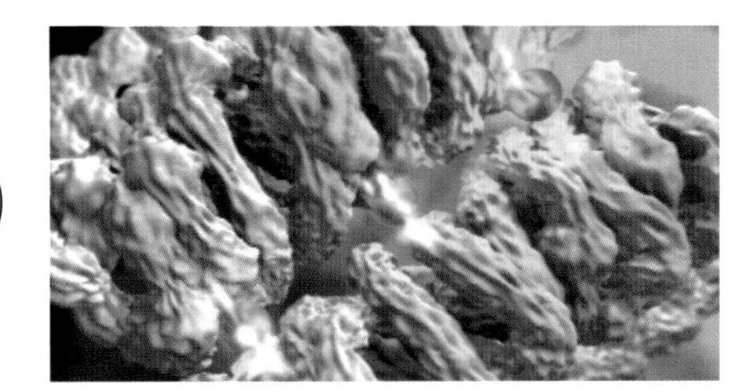

Index V14

Calcium ions (green) linking protein molecules (grey)

The calcium ion content of water can affect the colour of cooked vegetables as well as their texture, but indirectly, by its effect on pectin molecules. Since most vegetables require some softening during cooking, cooking in hard water means that longer is needed to achieve the optimum softening. During this longer cooking time, more chlorophyll is converted to phenophytin and the colour of green vegetables becomes browner.

Practical work

Two pieces of practical work are suggested. The first, *The effect of salt and of sodium hydrogencarbonate on the colour of vegetables*, is a relatively open-ended investigation. The second, *The effect of calcium ions on the texture of vegetables* is rather more prescriptive in terms of method, although it could be opened out if time allows.

The effect of salt and of sodium hydrogencarbonate on the colour of vegetables

Students are asked to devise and carry out experiments to investigate the effect of adding salt (sodium chloride) and sodium bicarbonate (sodium hydrogencarbonate) on the colour of cooked vegetables. There are many variables involved including concentration of additives, cooking time and type of vegetable. Peas or green beans are suggested because they can be cooked in small quantities and they have a relatively uniform colour. Any colour changes will have to be assessed with the naked eye and will inevitably be subjective. They could be made more objective by using a variation of the 'odd one out' test used for flavour. A tester is offered three samples to assess, one cooked in one way and two cooked in another. The tester is asked to pick the odd one out.

During trials, frozen peas were cooked in tap water and in water containing sodium hydrogencarbonate (approximately 40 g dm^{-3}). After boiling for five minutes, the peas in the sodium hydrogencarbonate solution were consistently rated by a tester as greener in colour than those in tap water, although the colour difference was small. The peas cooked in the alkaline solution were noticeably soggier than those cooked in tap water. As a further check, peas were cooked in tap water and in water containing a little vinegar to increase the concentration of H^+ ions. Those cooked in the acidic solution almost instantly acquired a very distinct yellow-green hue.

Apparatus and equipment

Each group of students will need:

- saucepan or large beaker in which to cook the vegetables
- spoon or large spatula to remove vegetables from boiling water
- access to a top pan balance
- access to a cooker or other means of boiling vegetables (*eg* Bunsen burner, tripod, gauze and heatproof mat)
- saucers (ideally white) or petri dishes on which to display samples for testing
- a tea strainer or similar would be useful for draining the vegetables.

Chemicals

Each group of students will need:

- sodium hydrogencarbonate
- sodium chloride (table salt)
- peas – fresh or frozen (or other green vegetables)
- vinegar (for the extension activity).

The effect of calcium ions on the texture of vegetables

This can be investigated by cooking dried peas that have been soaked overnight in (a) deionised water, (b) tap water, (c) 2% calcium chloride solution, (d) 4% calcium chloride solution. The peas are drained and then simmered in a beaker of water covered with a clock glass (to reduce evaporation) for 10 minutes. After this time a sample of peas is tasted to assess the texture. Sampling and tasting is continued at regular intervals and the time for each type of pea to achieve an acceptable texture is noted. Fuller details are given in

Revised Food Science, a Special Study, London: Longman, 1984.

Note. The Nuffield Chemistry Special Study materials are in the course of being made available for free download at ***www.chemistry-react.org*** *(accessed January 2005)*

Apparatus and equipment

Each group of students will need:

- saucepan with lid or large beaker with clock glass in which to cook the vegetables
- spoon or large spatula to remove vegetables from boiling water
- access to top pan balance
- access to cooker or other means of boiling the vegetables (*eg* Bunsen burner, tripod, gauze and heatproof mat)
- saucers or petri dishes on which to display samples for testing
- a tea strainer or similar would be useful for draining the vegetables.

Chemicals

Each group of students will need:

- dried peas
- deionized water
- 2% calcium chloride solution
- 4% calcium chloride solution.

To save time, some teachers may prefer to supply students with peas that have already been soaked.

Safety

- Wear eye protection.
- Take care with boiling water.
- Make sure that students follow proper hygiene precautions (see the section ***Experiments with food*** in ***How to use this material*** on page vii).
- Your employer's risk assessment should be consulted before carrying out this activity. This activity is covered by model (general) risk assessments widely adopted for use in UK schools and colleges such as those provided by CLEAPSS, SSERC, ASE and DfES. Bear in mind, however, that these may need some modification to suit local conditions.

Questions for students

These can be tackled after reading the material (above and repeated in the student's sheet) about the factors that affect the colour and texture of vegetables during cooking.

1. Look at the formula of a section of the pectin molecule. What will happen to some of the –COOH groups in water? How will this be affected by making the water more alkaline? Suggest how a Ca^{2+} ion could form a link between two pectin molecules. Is this link more likely to form in acidic or alkaline conditions?

2. Pectin forms water-retaining gels. Look at the formula of pectin. What features of the molecule allow it to bond with water molecules? What type of bonds are formed?

3. The pectin molecule contains –COOH groups. How will these react with sodium hydrogencarbonate? Write an equation for this reaction representing pectin as RCOOH. Would you expect the organic product of this reaction to be more or less soluble in water than pectin itself? Explain your answer. What would you predict to be the effect on the texture of vegetables of cooking them in water to which sodium hydrogencarbonate has been added?

Answers

1. They will lose a H^+ ion to form $-COO^-$. This effect will be more pronounced in an alkaline solution because the H^+ ions will react with OH^- ions to form H_2O and this will increase the dissociation of –COOH. The Ca^{2+} ion could form an ionic bond with two $-COO^-$ ions on different pectin molecules, thus linking them together. This is more likely in alkaline conditions as there will be more $-COO^-$ groups.

2. Pectin has a great many oxygen atoms which can form hydrogen bonds with water molecules. The hydrogen atoms of the –OH groups can also hydrogen bond with water molecules.

3. They will react together in a neutralisation reaction:

   $$RCOOH + NaHCO_3 \rightarrow RCOONa + CO_2 + H_2O$$

   The organic product will be more soluble because it is ionic, $RCOO^- Na^+$

   The texture would be softer because the pectins, which are responsible for the structure, would be more soluble.

Teacher's notes:

Kitchen**Chemistry**

Should beans be cooked with the lid on or off?

Should beans be cooked with the lid on or off?

Learning objectives

- To devise and carry out experiments to test simple hypotheses.
- To develop the idea of the 'fair test'.
- To learn about the effect of pressure on the boiling point of water.

Level

Age: 5–11.

Timing

One lesson of about one hour.

Description

Pupils devise and carry out experiments to investigate the suggestion that green vegetables will discolour if cooked with the lid on.

Teaching notes

The lesson could be started by introducing Heston Blumenthal as a 'scientific chef' and showing the video clip of him discussing the use of salt in cooking food and/or by discussing some of the issues listed below at a level appropriate for the pupils concerned.

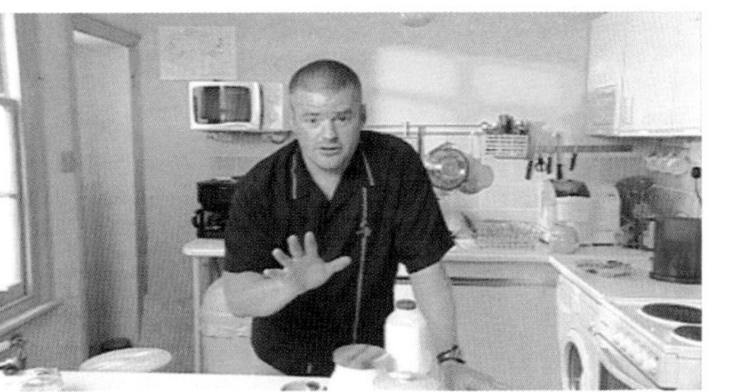

Index V01

Heston Blumenthal in the kitchen

One question posed by Heston Blumenthal early in his career as a 'scientific chef' was ***'Why do cooks add salt (sodium chloride) when cooking vegetables, for example green beans?'*** Possible reasons suggested by cooks included:

- it keeps the beans green
- it raises the boiling point of water so the beans cook faster
- it prevents the beans going soggy
- it improves the flavour.

A scientist colleague replied that there seemed to be no good reason because:

- only the acidity and calcium content of the water affect the colour of the beans
- adding salt ***does*** increase the boiling point of water but by such a small amount that it will make no difference to cooking times
- vegetables will go soggy if cooked for too long whether salt is added or not
- very little salt is actually absorbed onto the surface of a bean during cooking – typically 1/10 000 g of salt per bean which is too little to be tasted by most people.

Another piece of cooking folklore is that green vegetables will discolour if cooked with the lid on. Testing this could form the basis of a pupil investigation. The major difficulty will probably be in comparing the colours of different samples of vegetable – an 'odd one out' test might make this assessment more objective. Here the tester is offered three samples, one treated in one way and the other two in a different way, and has to try to pick the odd one out.

Some pupils may find the proving of a negative unsatisfying and teachers may wish to consider combining this investigation with one in which there *is* an effect and testing the effect of acidity/alkalinity or calcium ion content of the cooking water on colour, see ***What affects the colour and texture of cooked vegetables?*** (*page 35*) for some suggestions.

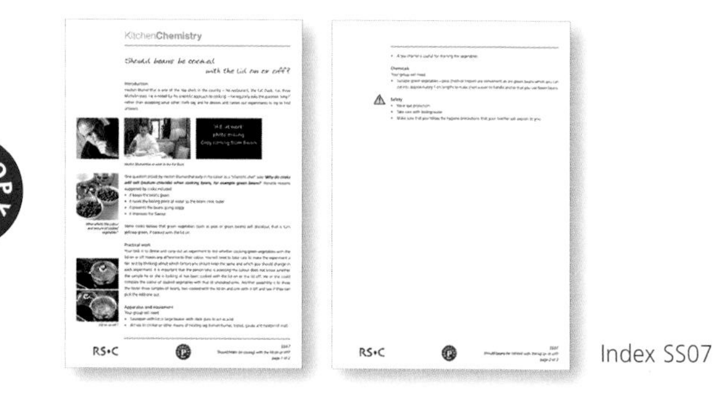

Index SS07

Practical work

The pupils are asked to devise and carry out an experiment to test whether cooking green vegetables with the lid on or off makes any difference to their colour. They will need to take care to make the experiment a fair test by thinking about which factors to keep the same and which to change in each experiment – cooking time, for example. It is important that the person who is assessing the colour does not know whether the sample he or she is looking at has been cooked with the lid on or the lid off. He or she could compare the colour of cooked vegetables with that of uncooked ones. Another possibility is to show the tester three samples of beans, two cooked with the lid on and one with it off and see if they can pick the odd one out.

Apparatus and equipment

Each group of pupils will need:

- saucepan with lid or large beaker with clock glass to act as a lid
- access to cooker or other means of heating (*eg* Bunsen burner, tripod, gauze and heatproof mat)
- a tea strainer is useful for draining the vegetables.

Chemicals

Each group of pupils will need:

- suitable green vegetables – peas (fresh or frozen) are convenient as are green beans which can be cut into approximately 1 cm lengths for convenience and economy.

Safety

- Wear eye protection.
- Take care with boiling water.

- Make sure that pupils follow proper hygiene precautions (see the section ***Experiments with food*** in ***How to use this material*** on page vii).
- Your employer's risk assessment should be consulted before carrying out this activity. This activity is covered by model (general) risk assessments widely adopted for use in UK schools and colleges such as those provided by CLEAPSS, SSERC, ASE and DfES. Bear in mind, however, that these may need some modification to suit local conditions.

What affects the colour and texture of cooked vegetables?

Further information

Perhaps not surprisingly, this piece of folklore is just that – lid on or lid off makes no difference. This was one of the first scientific tests carried out by Heston Blumenthal. With some groups of pupils it may be appropriate to discuss possible reasons for this myth having arisen. One may be a belief that the lid raises the pressure inside the pan and thus increases the boiling point of the water. Unless the pan is sealed (as in a pressure cooker, see below), there should be no pressure increase and thus no effect on the boiling point. This could easily be tested with a digital thermometer provided the saucepan lid has a hole through which the temperature probe will fit.

Factors that do affect the colour of green vegetables during cooking are (a) the acidity or alkalinity of the water (this is why some cooks add the weak alkali bicarbonate of soda (sodium hydrogencarbonate) to the water when cooking green vegetables), (b) the hardness (strictly the calcium content) of the water. For more details, see ***What affects the colour and texture of cooked vegetables?*** *(page 35)*.

The pressure cooker

The temperature at which a liquid boils increases as the external pressure increases. This is because boiling occurs when the vapour pressure of water is equal to the external pressure which means that bubbles can form in the body of the liquid. At normal atmospheric pressure, 100 kPa, water boils at 100 °C, but at twice this pressure it boils at about 125 °C. Since cooking is a series of chemical reactions and many chemical reactions roughly double in rate for every 10 °C temperature rise, this means that cooking under pressure is quicker. In fact for a 25 °C rise in temperature, food will cook over four times more quickly.

A pressure cooker has a sealed lid (with a valve to ensure that the pressure does not get too high). As the water boils, steam is produced and this raises the pressure inside the pan and increases the boiling point of the water. Visit this website to find out more about pressure cooking: *http://missvickie.com/workshop/howdoesit.html (accessed January 2005)* (unfortunately the temperatures are given in °F). Most pressure cookers are set to operate at 100 kPa above atmospheric pressure, *ie* at double the atmospheric pressure.

Mountaineers boiling water at a height of 6170 metres

Questions for pupils

1. Mountaineers often complain that it is difficult to make a decent cup of tea with water boiled at high altitude. Suggest why this problem might occur.

   This question is difficult. It could be set to pupils of appropriate ability if the pressure cooker and how it works has been discussed. Alternatively it could be used as part of a teacher-led discussion with the class.

2. Is it possible to cook a boiled egg at the summit of Mount Everest?

This question is also difficult but could be discussed with groups of appropriate ability.

Answers

1. Water boils at a higher temperature when the external pressure is raised and at a lower temperature when the pressure is reduced. High on a mountain, the atmospheric pressure is less than at sea level and the boiling point is reduced. This will affect the process of extracting soluble flavour components from the tea leaves into the aqueous solution that is the tea. The table gives some examples.

Mountain	Height / m	Atmospheric pressure / kPa	Boiling point of water / °C
Everest	8848	31	71
Mont Blanc	4807	55	85
Ben Nevis	1392	86	95

2. The answer depends on the coagulation temperature of the proteins in egg. One source quotes the coagulation temperature of albumen (a protein in egg white) as approximately 71 °C, the same as the boiling point of water at this altitude.

Teacher's notes:

Kitchen**Chemistry**

The chemistry of baking powder

RS•C

The chemistry of baking powder

Learning objective

- To revise aspects of stoichiometry calculations, acid / base reactions and organic acids.

Level

Age: post-16.

Timing

Approximately half an hour.

Description

A passage of reading about the chemistry of baking powder is followed by questions that test the understanding of familiar chemistry in an unfamiliar context. The topics covered include stoichiometric calculations and acid-base chemistry including organic acids.

Teaching notes

The lesson could be introduced by showing the video clip of Heston Blumenthal talking about the use of salt in cooking food and/or by discussing the issues listed below.

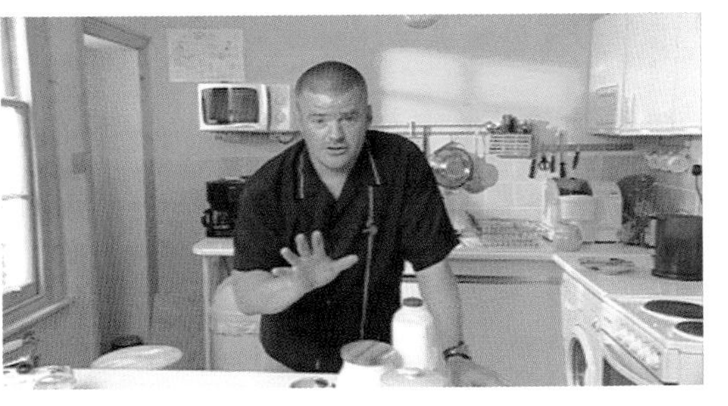

Index V01

Heston Blumenthal in the kitchen

The passage of student reading material and the questions are reproduced below.

KitchenChemistry

The chemistry of baking powder

[illegible]

RS•C

RS•C

RS•C

Index SS08

Baking powder

Sponge cakes made with (right) and without (left) baking powder

Many food products such as bread, sponge cakes and buns have a honeycomb structure which contains many bubbles. During cooking these bubbles are formed by a gas, and the mixture 'rises'. In some cases the gas is air which is whipped into the mixture before cooking and which expands during cooking. In other cases the gas is carbon dioxide. This can be formed either from the fermentation of sugar aided by yeast (as in making bread) or by chemicals that react to form carbon dioxide.

The most common chemical used for this purpose is sodium hydrogencarbonate, $NaHCO_3$ (more commonly called sodium bicarbonate, bicarbonate of soda or just 'bicarb'). This can form carbon dioxide in two ways:

- on heating

$$2NaHCO_3(s) \rightarrow Na_2CO_3(s) + CO_2(g) + H_2O(l)$$

- on reacting with an acid, such as hydrochloric acid (HCl)

$$NaHCO_3(s) + HCl(aq) \rightarrow NaCl(aq) + CO_2(g) + H_2O(l)$$

Cooks do not use a strong acid such as hydrochloric acid; they use instead a weak acid such as potassium hydrogentartrate (potassium hydrogen-2,3-dihydroxybutanedioate or potassium hydrogen-2,3-dihydroxysuccinate), also called cream of tartar. The formula of this acid is:

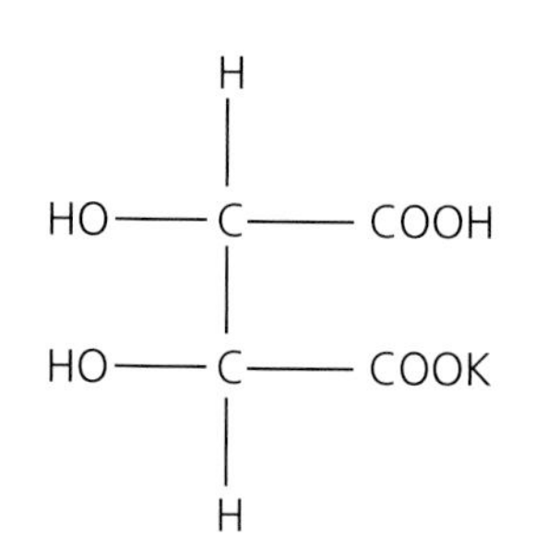

The gaseous product of the reaction is the same as with hydrochloric acid – carbon dioxide – but it is produced much more slowly. Potassium hydrogentartrate is a solid and this means that it is possible to mix it with the sodium hydrogencarbonate without the two reacting – they only react in the presence of water. This dry mixture is the basis of baking powder. The reaction is:

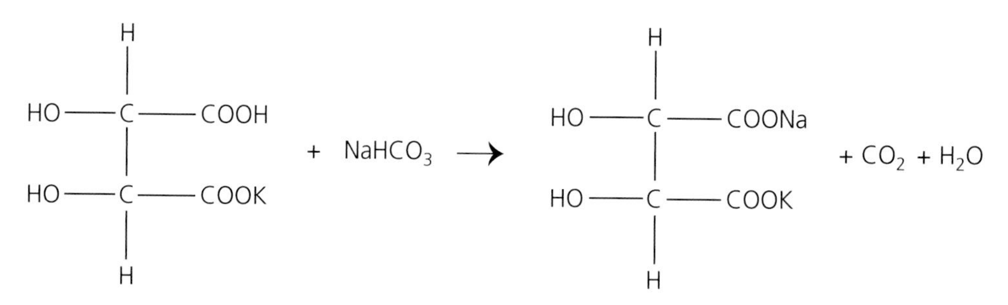

One problem with the use of potassium hydrogentartrate is that it is very soluble in water. So as soon as it becomes wet (when milk is added in a cake recipe, for example) it dissolves and reacts. This risks all the gas escaping while the cake mix is still liquid and before it goes in the oven. Most baking powders nowadays are so-called 'double acting'. This means that, along with the sodium hydrogencarbonate, they use a mixture of potassium hydrogentartrate and calcium dihydrogendiphosphate ($CaH_2P_2O_6$), which is also a solid acid. The potassium hydrogentartrate dissolves and reacts almost immediately (which makes the dish 'rise' on mixing) while the calcium dihydrogendiphosphate is slower to dissolve and will not react until the mixture is in the oven and the gas bubbles are trapped by the cake as it bakes.

Note. Both calcium dihydrogendiphosphate and potassium hydrogentartrate are acidic salts. The hydrogen atoms marked in red in the formula of calcium dihydrogendiphosphate below are acidic. That is, the salt can dissociate, losing the red hydrogen atoms as H^+ ions.

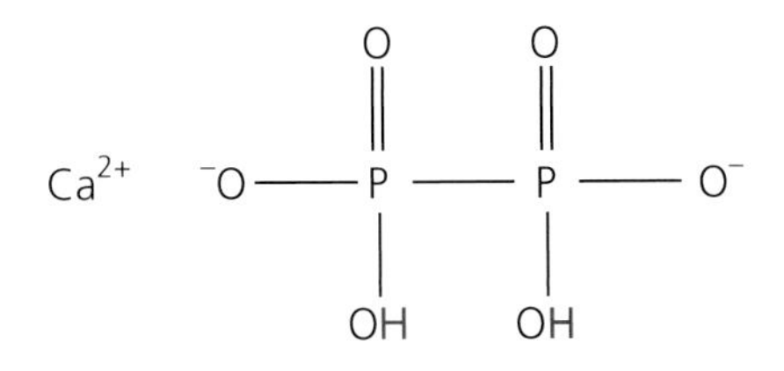

Questions

1. Baking powder will 'go off' if it is stored for some time in the kitchen, *ie* it will lose its ability to produce gas when moistened. Explain why this happens and suggest a method of storing the powder to minimise the problem.

2. Sodium hydrogencarbonate will decompose when heated, forming carbon dioxide and sodium carbonate:

   $2NaHCO_3(s) \rightarrow Na_2CO_3(s) + CO_2(g) + H_2O(l)$

   Sodium carbonate has a somewhat unpleasant 'alkaline' taste, so this method of decomposition must be avoided. Suggest how a manufacturer might formulate baking powder to prevent this problem occurring.

3. Calculate the relative molecular mass, M_r, of sodium hydrogencarbonate and of potassium hydrogentartrate. In what proportions by mass do they have to be mixed to be in the correct reacting ratio?

4. Mark the acidic hydrogen on a copy of the formula of potassium hydrogentartrate.

5. Estimate the radius and height in cm of a sponge cake. What is the volume of the cake in cm^3? (You will need to use the equation, volume of a cylinder radius r and height h = $\pi r^2 h$, but for this estimate it will be sufficient to take the value of π as 3.)

   One mole of any gas has a volume of 24 000 cm^3 at room conditions. If the cake were all gas, how many moles of gas would be present? What quantities of sodium hydrogencarbonate and of potassium hydrogentartrate would be required to produce this amount of gas?

6. As well as carbon dioxide, fermentation produces ethanol (C_2H_5OH, the alcohol found in alcoholic drinks). One simple sugar is glucose, $C_6H_{12}O_6$.

   Use this information to write a balanced equation for the fermentation of glucose.

   Bread rises because of fermentation. Suggest what happens to the ethanol produced when baking bread.

Use the balanced equation for fermentation and the balanced equation for the reaction of potassium hydrogentartrate with sodium hydrogen carbonate to compare the mass of baking powder with the mass of glucose required to produce the same volume of carbon dioxide.

7. Explain the meaning of the terms 'strong' and 'weak' as used to describe acids in the passage. How do these terms differ from the terms 'concentrated' and 'dilute'?

8. Explain why the hydrogen of the –COOH group in potassium hydrogentartrate is more acidic than the two –OH hydrogens.

Answers

1. Sodium hydrogencarbonate and potassium hydrogentartrate react together in the presence of water. The moisture in the air in a typical kitchen will be enough to bring this about over a period of time. The solution is to store baking powder in a tightly sealed container.

2. Ensure that there is sufficient acid in the formulation to react with all the sodium hydrogencarbonate so that none will be left to produce sodium carbonate.

3. M_r $NaHCO_3$ = 84, M_r $C_4H_5O_6K$ = 188. They react 1 : 1 so they must be mixed in the ratio 84 g to 188 g.

4. The hydrogen atom marked in red is acidic.

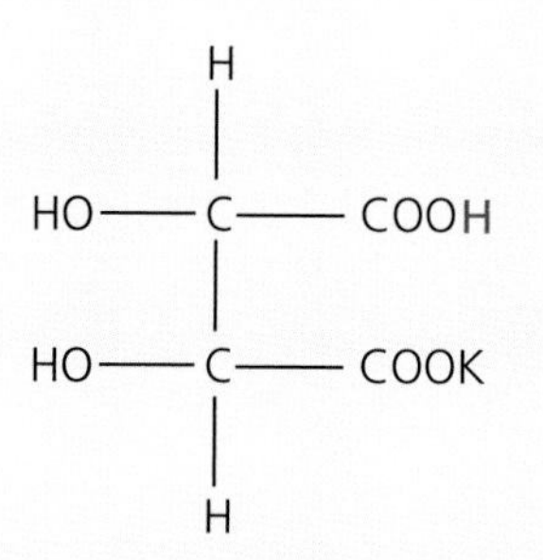

5. Various answers are acceptable. Look for sensible estimates of size.

   For example if h and r are both estimated as 10 cm, Volume = 3 x 10 x 10 x 10 = 3000 cm^3, which is 1/8 mol so 1/8 quantities calculated in question 3 would be required, *ie* approximately 10 g $NaHCO_3$ and 23g $C_4H_5O_6K$.

6. $C_6H_{12}O_6 \rightarrow 2C_2H_5OH + 2CO_2$

   The ethanol (boiling point 78 °C) evaporates at typical oven temperatures (over 200 °C).

To produce 1 mol carbon dioxide by fermentation requires 1/2 mol glucose (90 g).

To produce 1 mol carbon dioxide from baking powder requires a mixture of 1 mol sodium hydrogencarbonate and 1 mol potassium hydrogentartrate (see answer 3). This is a total of 272 g, *ie* almost three times as much.

7. Strong means fully dissociated into H^+ ions in solution, weak means only partially dissociated into H^+ ions in solution. Concentrated and dilute refer only to the number of moles of acid dissolved in each cubic decimetre of solution and not at all to the degree of dissociation.

8. The $–COO^-$ group left after loss of H^+ from –COOH has the charge delocalised over all three atoms which gives it increased stability. This does not occur with the $–O^-$ group left after loss of H^+ from an –OH group.

Further information

Further information on baking powder and links to some simple experiments can be found at *http://users.rcn.com/sue.interport/food/bakgsoda.html (accessed January 2005).*

Teacher's notes:

....

....

....

....

....

....

....

....

....

....

....

....

....

....

....

Kitchen**Chemistry**

The structure of ice and water

RS•C

The structure of ice and water

Learning objectives

- Revision of structure and bonding, hydrogen bonding, electronegativity, VSEPR, polarity and dipole moments.

Level

Age: post-16.

Timing

The demonstration takes a couple of minutes. Interpretation and discussion may take up to half an hour depending on the detail covered and the ability of the group. If the questions are to be attempted, this could take an extra half hour or more.

Iceberg drifting in the open ocean

Description

Using a freezer it is possible to make a block of ice with a hollow in it that can be filled with water. If this ice block is placed in a microwave oven for about 30 seconds, the water will be seen to boil while the ice remains frozen. This unexpected observation can lead to a discussion of the bonding and structure of water and of ice, suitable for post-16 students. The discussion can be extended to cover a number of topics – hydrogen bonding, giant structures, electronegativity, shapes of molecules, polarity, dipole moments – and could be used to spark off a stimulating revision lesson to cover some or all of these topics. Questions are supplied that could be attempted by students after the discussion, possibly as homework.

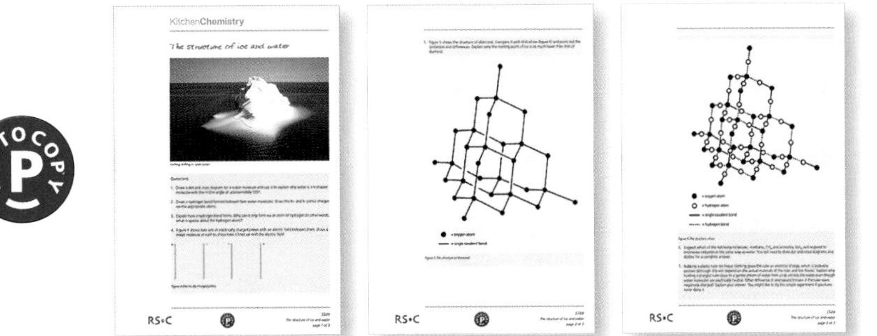

Index SS09

Teaching notes

The demonstration

It will be necessary to make the ice block before the lesson, keeping it in the freezer until required.

The easiest way to make the hollowed out block of ice is to take a small container such as a 400 cm^3 beaker part-filled with water and float a second container (such as a small, empty yoghurt pot) in it weighted down with sand or lead shot so that it floats upright, see Figure 1.

Place the beaker in the freezer taking care that the yoghurt pot remains in the centre of the beaker and does not float to the side. Leave until completely frozen through (this may take some hours). When frozen, remove from the freezer and pour the sand or lead shot out of the yoghurt pot. The yoghurt pot can then be part-filled with water. Leaving the pot in place reduces the tendency of the ice to be melted by heat conducted from the water but some teachers may prefer to remove it and pour the water directly into the hollow in the ice. (Teachers may feel that it is worth preparing more than one ice block in case a repeat demonstration is required.)

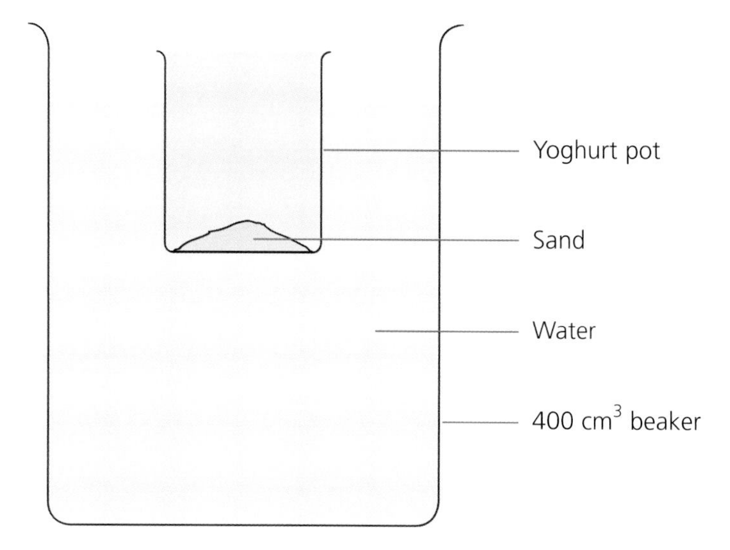

Figure 1 The beaker and yoghurt pot assembly

Place the beaker containing ice, yoghurt pot and water in the microwave oven on full power for the minimum time required for the water to boil (prior practice with the size of beaker and oven used will give a good idea of how long this will take). As soon as the water in the yoghurt pot begins to boil, remove the beaker as quickly as possible to show the audience that it is boiling. (It will be difficult for the audience to see inside the microwave oven.) A thermometer, preferably digital, may be used to confirm the water temperature.

Heat is, of course, conducted from the hot water to the ice, which will eventually melt, so leaving the experiment too long will reduce its impact. For the same reason it is important that the ice has been left long enough in the freezer to attain the temperature of the inside of the freezer (usually about -20 °C). Use the smallest yoghurt pot available as there is then less surface area for heat transfer from the boiling water to the ice.

Apparatus and equipment

You will need:

- one beaker about 400 cm^3
- one small plastic container to fit easily inside the beaker – a small yoghurt pot, for example
- a few grams of sand or lead shot
- thermometer (0–100 °C), preferably digital for ease of viewing by the audience
- access to a freezer
- access to a microwave oven.

Safety

- Wear eye protection.
- Take care with boiling water.
- Your employer's risk assessment should be consulted before carrying out this activity. This activity is covered by model (general) risk assessments widely adopted for use in UK schools and colleges such as those provided by CLEAPSS, SSERC, ASE and DfES. Bear in mind, however, that these may need some modification to suit local conditions.

Interpretation

It will first be necessary to explain to the students how microwave heating works.

The O-H bond in a water molecule has a dipole $O^{\delta-}–H^{\delta+}$ because the electronegativity of oxygen (3.5) is significantly greater than that of hydrogen (2.1). (The discussion of electronegativity in terms of shielded nuclear charge could also be brought up.) Because of its angular shape (also worth discussion in terms of valence shell electron pair repulsion) the water molecule has an overall dipole moment as shown in Figure 2.

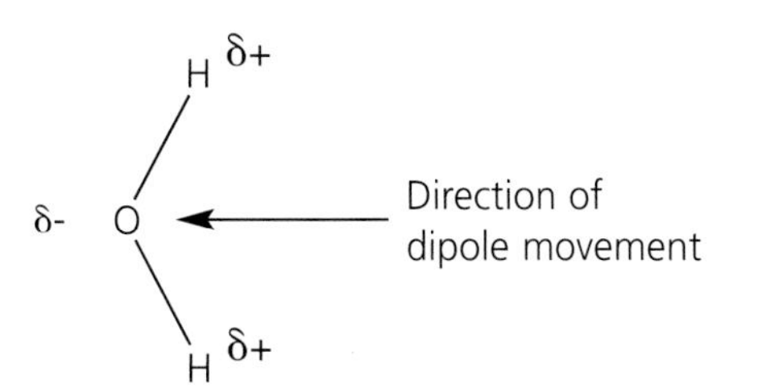

Figure 2 The water molecule's dipole moment

Microwave radiation, like all other sorts of electromagnetic radiation, consists of oscillating electric (and magnetic) fields. Microwaves used in household ovens have a frequency of 2.45 GHz *ie* 2.54 thousand million complete vibrations per second. This means that the electric field changes direction 4.9 thousand million times per second. Dipoles try to line up with electric fields with their δ- ends towards the positive and their δ+ ends towards the negative end. So water molecules try to flip their direction to keep up with the changes in the electric field; this makes the water molecules rotate. The rotating water molecules collide with other molecules and this makes them move from place to place. This movement is what we call heat.

In ice, hydrogen bonding between the water molecules (also worth discussion) leads to a lattice structure in which each oxygen atom has two covalent bonds to hydrogen atoms within the water molecule and two hydrogen bonds to hydrogen atoms in other water molecules. So each oxygen atom is surrounded tetrahedrally by four hydrogens and the overall structure (Figure 3) resembles that of diamond (see the figure in question 5, page 61).

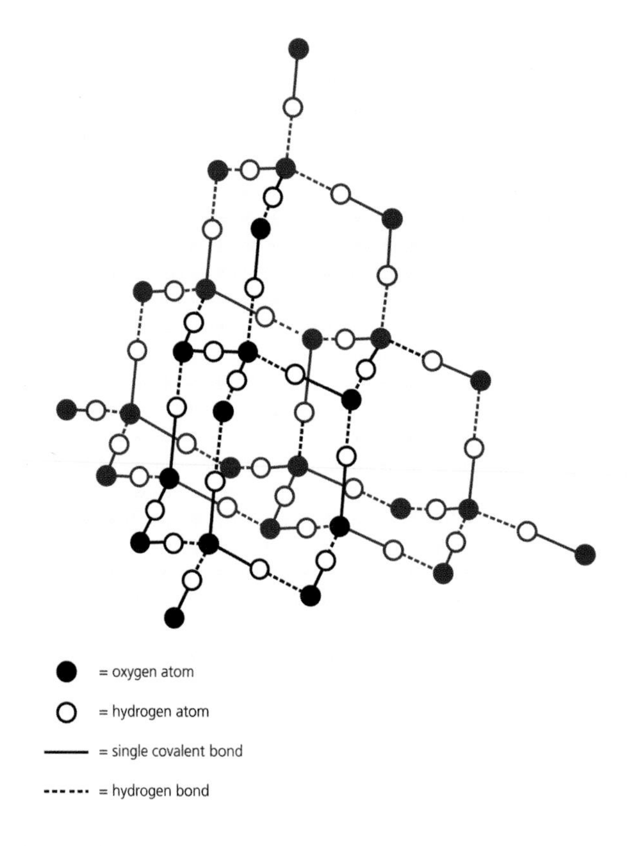

Figure 3 The structure of ice

NB Rotatable models of both the water molecule and the giant structure of ice are available on the CDROM and web versions of this material.

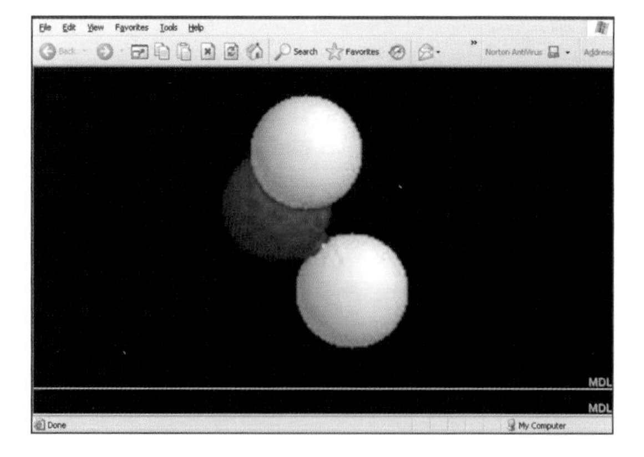

Water molecule

Index CM01

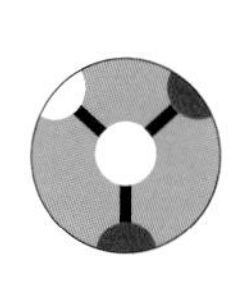

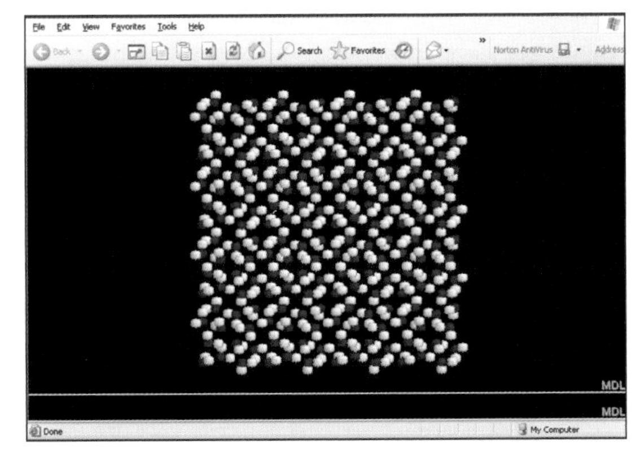

The giant structure of ice

Index CM02

So the water molecules in ice are not free to rotate and the ice does not heat up when subjected to microwave radiation.

Note that in this diamond-like structure, the water molecules are on average slightly further apart than they are in liquid water. So ice is slightly less dense than water and floats.

If looked at closely, there are several complications in the interpretation of this experiment and some of these could be discussed with able students.

- The ice and the water do not start at the same temperature. Ice from a typical freezer will be at approximately -20 °C while water from the tap will be at about 20 °C.
- Heat will be conducted from the water to the ice and so ice in contact with the water will tend to be melted by this heat as well as by any heat absorbed from the microwave radiation.
- The specific heat capacities of water and ice are not the same. That of water is approximately 4.2 $J\ g^{-1}\ {}^{\circ}C^{-1}$, and that of ice is approximately 2 $J\ g^{-1}\ {}^{\circ}C^{-1}$.
  Thus the temperature of ice would be expected to increase about twice as fast as that of water for the same amount of heat absorbed.
- The heat generated inside a microwave oven is not evenly distributed. There are 'hotspots'.

The questions below are reproduced in the student's worksheet.

Questions

1. Draw a dot and cross diagram for a water molecule and use it to explain why water is a V-shaped molecule with the H-O-H angle of approximately 105°.
2. Draw a hydrogen bond formed between two water molecules. Draw the $\delta+$ and $\delta-$ partial charges on the appropriate atoms.

3. Explain how a hydrogen bond forms. Why can it only form via an atom of hydrogen (in other words, what is special about the hydrogen atom)?

4. The figure shows two sets of electrically charged plates with an electric field between them. Draw a water molecule in each to show how it lines up with the electric field.

Electrically charged plates

5. The figure shows the structure of diamond. Compare it with that of ice shown on the next page and point out the similarities and differences. Explain why the melting point of ice is so much lower than that of diamond.

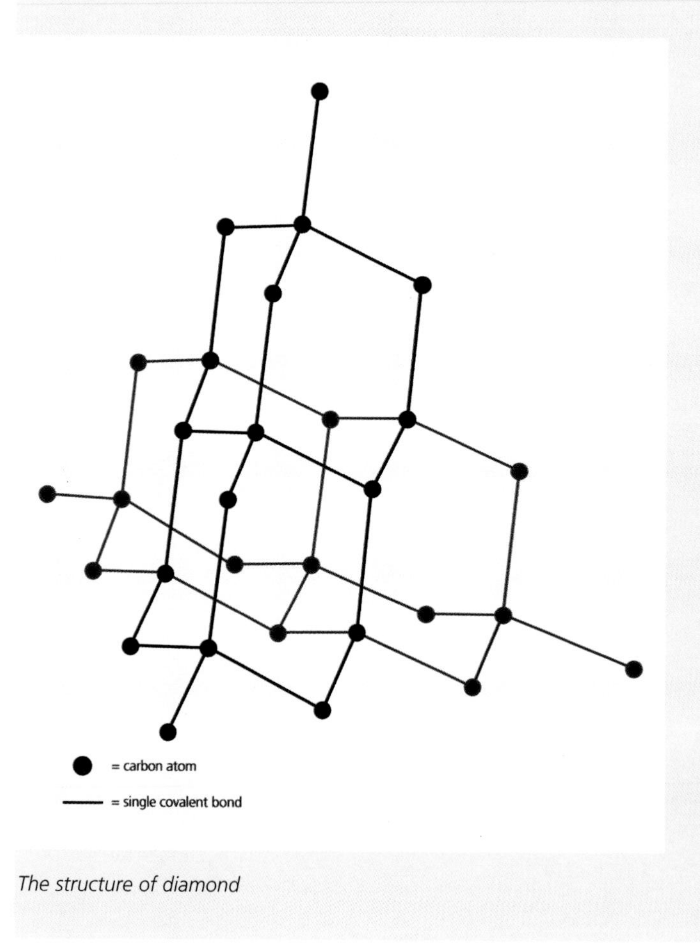

The structure of diamond

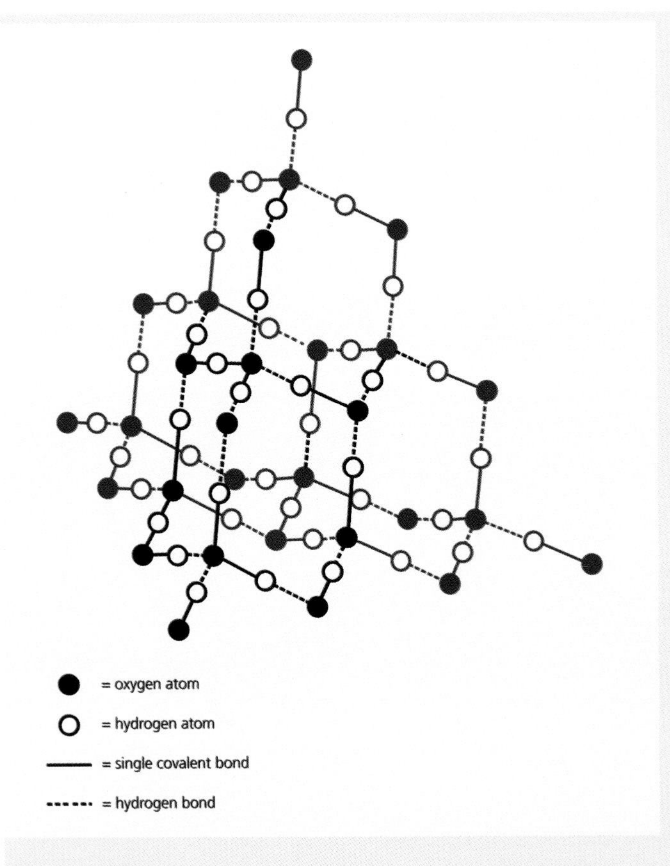

The structure of ice

6. Suggest which of the following molecules: methane, CH_4, and ammonia, NH_3, will respond to microwave radiation in the same way as water. You will need to draw dot and cross diagrams and dipoles for a complete answer.

7. Rubbing a plastic ruler on fleece clothing gives the ruler an electrical charge, which is probably positive (although this will depend on the actual materials of the ruler and the fleece). Explain why holding a charged ruler close to a gentle stream of water from a tap attracts the water even though water molecules are electrically neutral. What difference (if any) would it make if the ruler were negatively charged? Explain your answer. You might like to try this simple experiment if you have never done it.

Further information

In ice, the O–H covalent bond length is 0.096 nm and the O---H hydrogen bond length is 0.180 nm – figures that might prove useful in discussion. For simplicity, no attempt has been made to indicate this in the figures showing the structure of ice.

NB Rotatable models of the giant structure of ice and diamond are given on the CDROM and web version of this material.

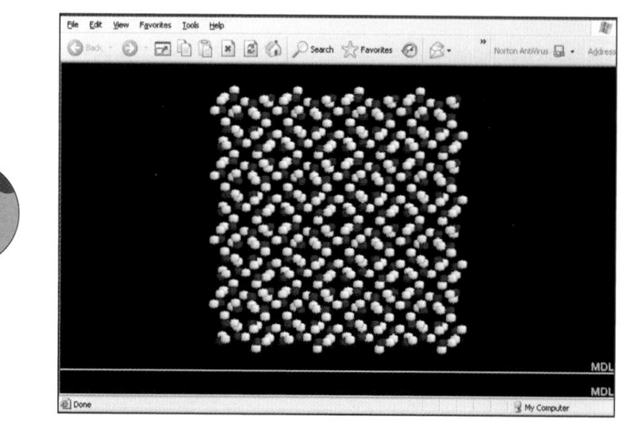

The giant structure of ice
Index CM02

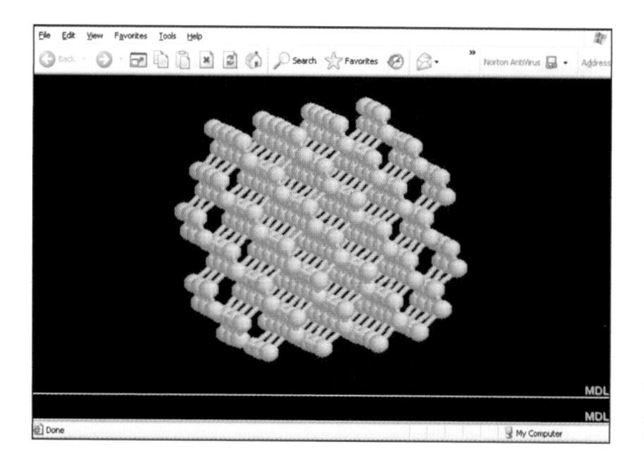

The structure of diamond
Index CM03

Answers

1. In water, the oxygen atom has two shared and two lone pairs of electrons in its outer shell. The mutual repulsion of these electron pairs leads to a shape based on a tetrahedron but with two points missing, *ie* V-shaped. The lone pairs are closer to the oxygen nucleus than the shared pairs and repel more effectively thus 'squeezing down' the bond angle from 109.5° to approximately 105°.

2. 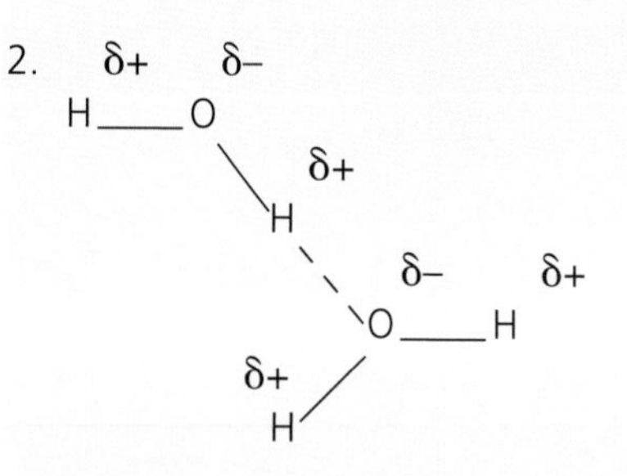

3. The hydrogen atoms in a water molecule each acquire a partial positive charge and the oxygen atom a partial negative one because the pair of electrons that bonds them 'feels' a larger shielded nuclear charge from the oxygen nucleus than from the hydrogen one (in other words, oxygen is more electronegative than hydrogen). Thus the hydrogen and the oxygen in adjacent water molecules are attracted electrostatically. However the hydrogen bond is more than just electrostatic. Because of the large unshielded charge (high electronegativity) of the oxygen atom, the hydrogens are

almost denuded of electrons and can receive a lone pair from the oxygen atom of another water molecule to form what is in effect a dative covalent bond – the hydrogen bond. The small size of the hydrogen atom means it has an intense electric field close to it and it is this that strongly attracts the oxygen lone pair.

4. 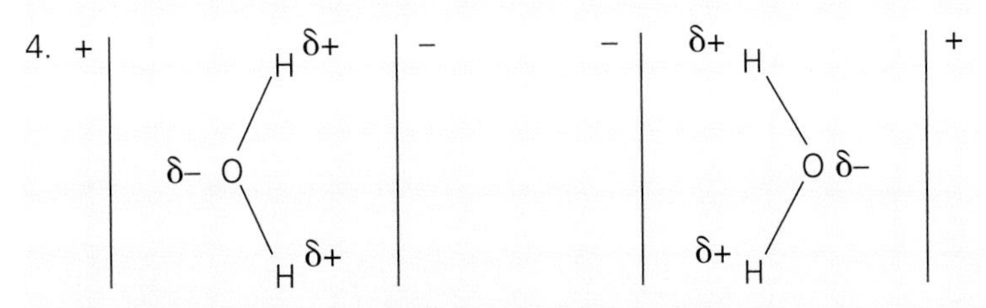

5. In diamond, each carbon atom is surrounded tetrahedrally by four other carbon atoms held by covalent bonds. In water, each oxygen atom is surrounded tetrahedrally by four other oxygen atoms but with hydrogen atoms in between. Two of these hydrogens are covalently bonded and two hydrogen-bonded.

   To melt ice it is necessary to break only the hydrogen bonds. To melt diamond, covalent bonds (which are much stronger) must be broken.

6. Ammonia has an overall dipole moment due to its pyramidal shape while methane has no overall dipole moment – its tetrahedral shape means that the individual bond dipoles (which are small anyway) cancel. So in microwave radiation, ammonia molecules will tend to rotate and methane ones will not.

7. The water molecules will 'flip' so that the δ- ends of their dipoles are towards the positively charged ruler and the δ+ ends away from it. The δ- ends are closer to the rod than the δ+ ones and are attracted to the positively charged ruler. (Note that since the water molecule is neutral and there are two δ+ to each δ-, the δ- is twice as large as the δ+.) In the case of a negatively charged ruler, the water molecules will 'flip' so that the δ+ ends of their dipoles are towards the ruler and they will still be attracted.

Teacher's notes:

....

....

....

....

....

....

....

Kitchen**Chemistry**

Why do pans stick?

Why do pans stick?

Learning objective

- Revision of aspects of the chemistry of carbohydrates, fats, proteins, and of the processes of cracking and polymerisation.

Level

Age: post-16.

Timing

Approximately half an hour.

Description

A passage of reading about the chemistry that causes food to stick to pans during cooking is followed by questions that test the understanding of familiar chemistry in an unfamiliar context. The topics covered include proteins, fats, carbohydrates, cracking and polymerisation.

Teaching notes

The exercise could be used for revision, as a homework or in case of teacher absence. It could be introduced by discussing Heston Blumenthal as a 'scientific chef' who needs to understand the chemistry of the cooking process in order to become a better cook and devise new and exciting recipes. The passage from the student's worksheet is reproduced here along with the answers to the questions.

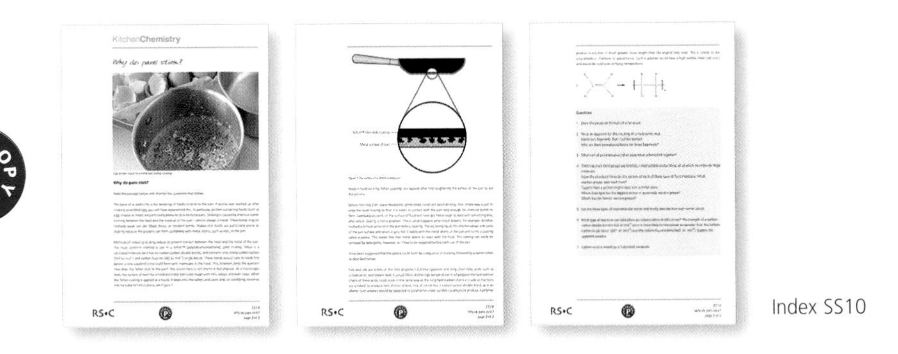

Index SS10

Why do pans stick?

The bane of a cook's life is the tendency of foods to stick to the pan. If you've ever washed up after cooking scrambled egg, you will have experienced this. Protein-containing foods (such as egg, cheese or meat) are particularly prone to stick to metal pans. Sticking is caused by chemical bonds forming between the food and the material of the pan – almost always a metal. These bonds may be relatively weak intermolecular forces or covalent bonds. Protein-rich foods are particularly prone to sticking because the proteins can form complexes with metal atoms, such as iron, in the pan.

Methods of reducing sticking reduce or prevent contact between the food and the metal of the pan. The most common method is the use of a Teflon® (poly(tetrafluoroethene), ptfe) coating. Teflon® is a saturated molecule and contains only strong carbon-carbon (347 kJ mol^{-1}) and carbon-fluorine (485 kJ mol^{-1}) single bonds. These bonds would have to break first before a new covalent bond could form with molecules in the food.

This, however, begs the question, 'how does the Teflon® stick to the pan?' The reason here is not chemical but physical. At a microscopic level, the surface of even the smoothest metal pan looks rough with hills, valleys and even caves. When the Teflon® coating is applied as a liquid, it seeps into the valleys and caves and, on solidifying, becomes mechanically locked in place, see Figure 1.

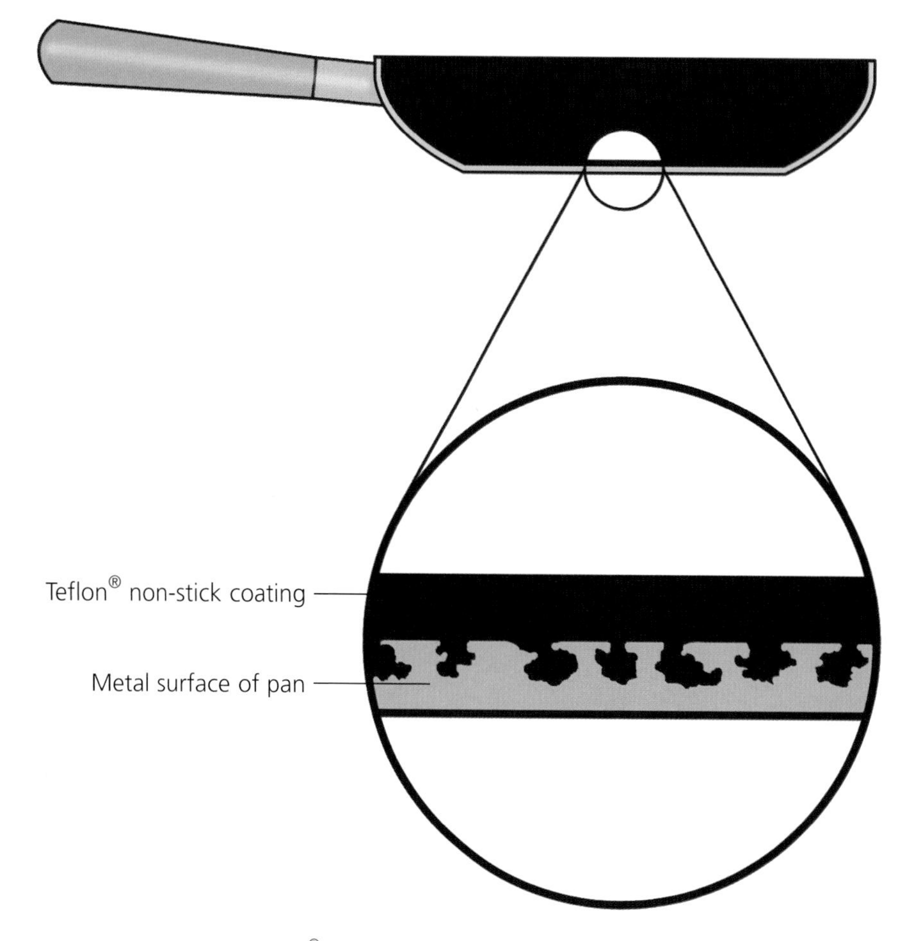

Figure 1 The surface of a Teflon® coated pan

Modern hard-wearing Teflon® coatings are applied after first roughening the metal surface of the pan to aid this process.

Before non-stick pans were developed, good cooks could still avoid sticking. One simple way to do this is just to keep the food moving so that it is never in contact with the pan long enough for chemical bonds to form. Eventually the proteins on the surface of the food will have got hot enough to react with something else, after which sticking is not a problem. This is what happens when meat browns, for example. Another method is to heat some oil in the pan before cooking. The oil, being liquid, fills in the valleys and caves of the pan surface and when it gets hot it reacts with the metal atoms of the pan and forms a coating called a patina. This leaves few free metal atoms to react with the food. This coating can easily be removed by detergents, however, so it has to be reapplied before each use of the pan.

It has been suggested that the patina could form by a sequence of cracking followed by polymerisation as described below.

Why do pans stick?

Fats and oils are esters of the triol propane-1,2,3-triol (glycerol) and long chain fatty acids such as octadecanoic acid (stearic acid, $CH_3(CH_2)_{16}COOH$). At the high temperatures reached in a frying pan the hydrocarbon chains of these acids could crack (in the same way as the long hydrocarbon chains in crude oil fractions are cracked) to produce two shorter chains, one of which has a carbon-carbon double bond, *ie* is an alkene. Such alkenes would be expected to polymerise under suitable conditions to produce a polymer of much greater chain length than the original fatty acids. This is similar to the polymerisation of ethene to poly(ethene). Such a polymer would have a high relative molecular mass and would be solid even at frying temperatures. The equation shows one example of such a polymerisation reaction where R and R′ would each represent a variety of different groups.

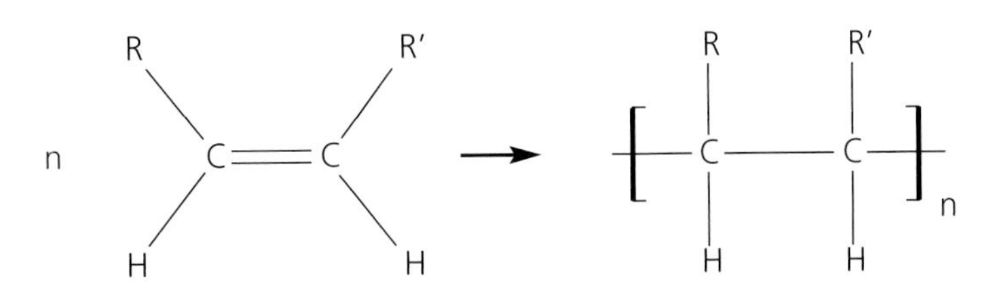

Questions

1. Draw the structural formula of a fat or oil.

2. Write an equation for the cracking of octadecanoic acid. Name two fragments that could be formed. Why are there several possibilities for these fragments?

3. What sort of polymerisation takes pace when alkenes link together?

4. The three main food groups are fats/oils, carbohydrates and proteins, all of which are relatively large molecules. Draw the structural formula of a section of each of these types of food molecules. What reactive groups does each have? Suggest how a protein might react with a metal atom. Which food type has the biggest variety of potentially reactive groups? Which has the fewest reactive groups?

5. List the three types of intermolecular forces in order of their strength (strongest first) and briefly describe how each comes about.

6. What type of reactions can take place at carbon-carbon double bonds? The strength of a carbon-carbon double bond is 612 kJ mol^{-1} yet it is more likely to be involved in reactions than the carbon-carbon single bond (347 kJ mol^{-1}) and the carbon-fluorine bond (485 kJ mol^{-1}). Explain this apparent paradox.

7. Explain what is meant by a 'saturated' molecule.

Answers

1.

The length of the side chains can vary.

2. For example

$CH_3(CH_2)_{16}COOH \rightarrow CH_3(CH_2)_7CH{=}CH_2 + CH_3(CH_2)_6COOH$

The fragments are dec-1-ene and octanoic acid.

The hydrocarbon chain could break at any point and the carbon-carbon double bond could be in either fragment so there are many acceptable answers.

3. Addition polymerisation.

4.

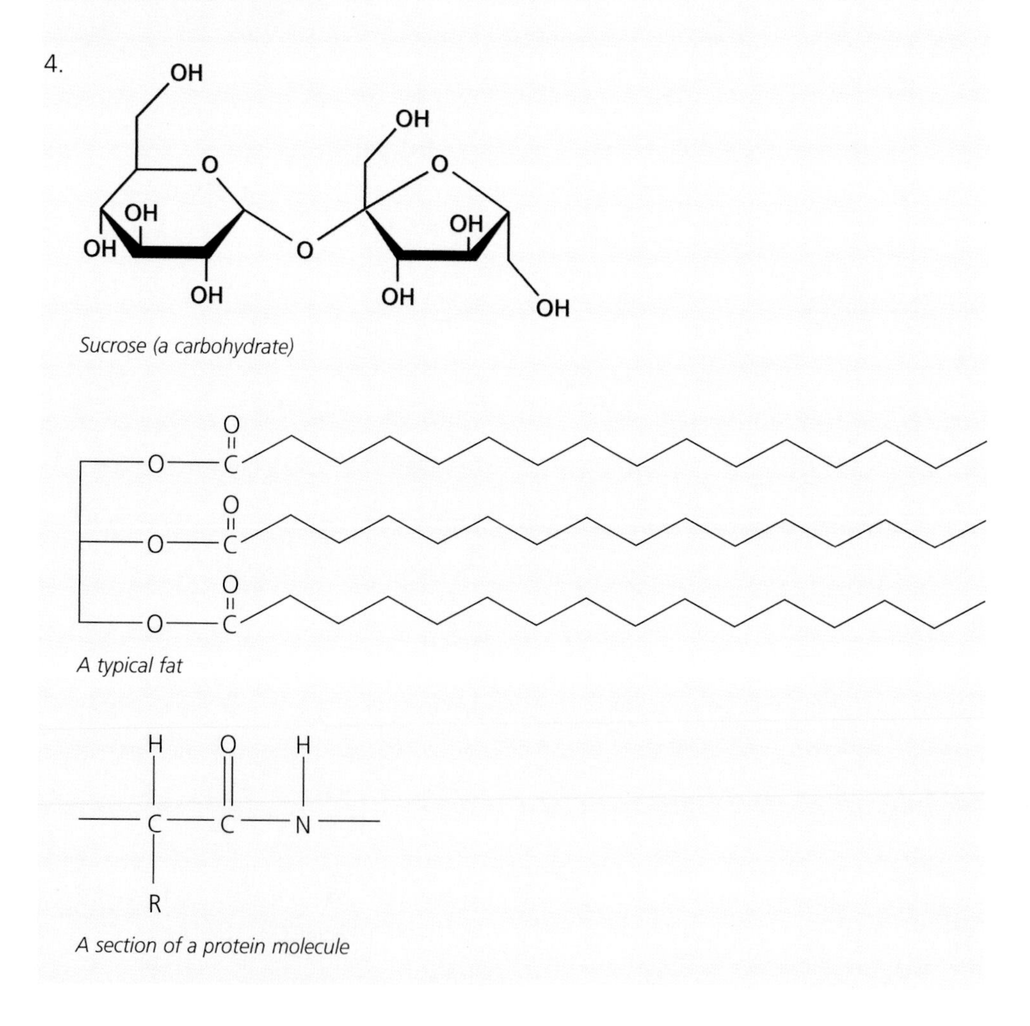

Sucrose (a carbohydrate)

A typical fat

A section of a protein molecule

Carbohydrates have –OH (alcohol) and R-O-R (ether) groups.
Fats have ester groups.

Proteins have amide groups and a number of other reactive groups incorporated as part of the side chain (R).

A protein could form a complex with a metal atom by donating the lone pair of its NH group into a part-full electron shell on the metal atom.

5. Hydrogen bonds, dipole-dipole interactions and van der Waals interactions.

   Hydrogen bonds form between a highly electronegative atom (F, O or N) and a hydrogen atom that is covalently bonded to another highly electronegative atom.

   Dipole-dipole interactions form between an atom that has a permanent $\delta+$ charge and one that has a permanent $\delta-$ charge.

   Van der Waals interactions form between all atoms as a result of instantaneous dipole-induced dipole attraction.

6. Addition reactions.

   A carbon-carbon double bond has two parts – a σ bond and a π bond. In addition reactions, the π bond (which is weaker) breaks and the σ bond remains intact. In fact the strength of the π bond is approximately 612 - 347 kJ mol^{-1} = 265 kJ mol^{-1}.

7. A saturated molecule has no multiple bonds.

Teacher's notes:

....

....

....

....

....

....

....

....

....

....

Kitchen**Chemistry**

Enzymes and jellies

Learning objectives

- To devise and carry out experiments to test simple hypotheses.
- Revision of enzymes.

Level

Age: This material could be tackled at a number of levels of sophistication making it suitable for a range of age groups from 7 to post-16.

Timing

Approximately one hour depending on what is attempted. It may be necessary to leave some experiments to be examined the following lesson to note the results.

Description

Party jelly will not set if it contains fresh pineapple but it will do so if tinned pineapple is used. Students investigate this observation with a view to finding an explanation. The explanation is that pineapple contains a proteolytic enzyme that breaks down gelatine, the setting agent.

Teaching notes

The lesson could be started by introducing Heston Blumenthal as a 'scientific chef' and showing one or other of the video clips illustrated below. In the first, Heston discusses jellies and gels in general, and in the second he demonstrates the action of the enzyme in pineapple on the mouth and also shows how to make a pineapple jelly by using chillies to denature the enzyme. Which clip to show depends on how the teacher envisages the lesson being structured and the age group of the students.

Index V02

Commercial fruit jelly in its 'raw' form

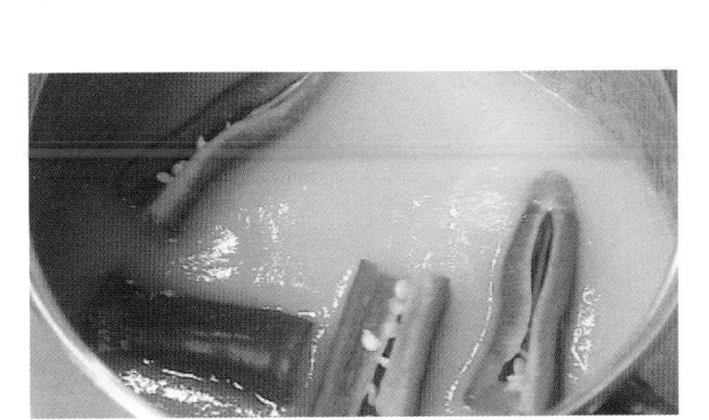

Index V03

Chillies added to fresh pineapple jelly mixture

Alternatively the teacher could start things off by drawing attention to the warning on jelly packets that '**adding fresh pineapple, kiwi or papaya fruit will prevent the jelly from setting**'.

Some fruits prevent jelly from setting

The lesson could be developed by a teacher demonstration, making jelly from a packet bought at the supermarket (a) by following instructions on the pack (b) by substituting some of the water in the recipe with fresh pineapple puree and (c) substituting some of the water in the recipe with tinned pineapple puree. Access to a freezer will speed up the setting process but it may well be more convenient to show samples made earlier.

Party jellies are based on gelatine, which is derived from the protein collagen. Collagen is based on three intertwined α-helices – when these are treated with hot water, acids or alkalis, the helices unravel and a three-dimensional partly cross-linked structure is formed. This is gelatine. When a warm, aqueous solution of gelatine sets, it forms a three-dimensional matrix that traps water – the jelly. Fresh pineapple contains a protease enzyme called bromelin that catalyses the breakdown (hydrolysis, *ie* reaction with water) of collagen so that adding fresh pineapple to the jelly solution prevents it from setting. Tinned pineapple has, like all tinned foods, been heated to a high temperature to destroy microorganisms. This denatures the enzyme so the jelly will set normally if tinned pineapple is added.

Purees of fresh and of canned pineapple made in a food processor are convenient for this experiment. Frozen pineapple may be substituted for fresh as the freezing process does not denature the enzyme.

Interestingly, fresh pineapple jellies can be made to set by adding chilli – the chilli (capsicum) contains a chemical that destroys the enzyme that prevents the jelly setting. This is the basis of a dish served at Heston Blumenthal's restaurant, the Fat Duck. The chefs at the Fat Duck have, however, found that this works only about 70% of the time. Students could be told this and asked to suggest reasons. Suggestions might include that different chillies contain different amounts of the chemical (see also ***How hot are chilli peppers?*** on page 115), length of cooking time, temperature *etc*.

Younger age groups will find plenty of opportunities in this activity to control variables (quantities, temperatures *etc*) while older students may be able to focus more on the process of denaturing the enzyme (how long does this take at different temperatures, for example?).

Method

No specific instructions are given as the investigation is essentially open-ended and will also vary according to age, ability and prior knowledge of the students (whether they know about enzymes and that they are denatured by heat, for example). What follows are some general indications of the method and some suggestions as to what to vary. The student's worksheet gives similarly general points rather than specific instructions.

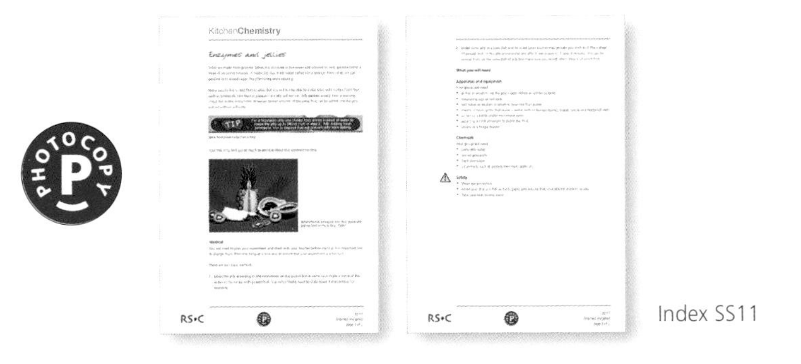

Index SS11

Making the jelly

Jelly can be bought in a variety of flavours in packs each containing a block of 12 squares. It can be made up as per the instructions on the pack using boiling water or microwave heating. The jelly is first dissolved in boiling water; cold water is then added and the jelly sets as it cools. Setting of the jelly can be speeded up by using a mixture of ice cubes and cold water in place of the cold water and/or by placing the jelly in the fridge or freezer. Students could scale down the quantities and work on single squares of jelly but this has the disadvantage that the water cools very quickly and may not easily dissolve the jelly. Teachers may prefer to make up the liquid jelly in bulk, keeping it warm on a water bath, and students can use portions of this as required. Colourless, unflavoured gelatine is available in shops but is more expensive than party jelly.

Variables

The simplest experiment is to make up jelly as per the instructions on the packet and time how long it takes to set (this may be as little as 10 minutes for about 25 cm^3 of jelly in a petri dish in a freezer). Replacing some of the water in the recipe with an equivalent quantity of pureed pineapple will cause the jelly not to set.

Things to try include:

- What effect does pureed tinned pineapple have?
- What effects do other fruit purees have – *eg* kiwi fruit, papaya, apple *etc*?
- For those fruits that affect setting, what is the effect of heating them first (in boiling water, for example)?
- What is the effect of adding chillies to those fruits that affect setting?
- Does the amount of pineapple puree added make any difference, *ie* is there a level below which the effect on setting is reduced or non-existent?
- What is the effect of heating the fresh pineapple puree – how long and to what temperature must it be heated before the jelly will set?

What effect do pineapple, kiwi fruit, guava and papaya have on the setting of jelly?

Alternative method

An alternative method is for students to make (or be supplied with) already-set jelly in petri dishes. Drops of fruit puree can be placed on the jelly and the jelly examined after some time – about 30 minutes. With pineapple, kiwi fruit and, to a lesser extent, papaya a small indentation appears under the drop after several minutes showing that the gel has broken down. This method is economical, as several tests can be carried out on a single petri dish of jelly. If using this method, it may be worth making the jelly with a little less water than usual so that the jelly is firmer. It may be necessary to leave the petri dishes to be examined during the following lesson to see clear-cut results.

Apparatus and equipment

Each group of students will need:

- dishes in which to set the jelly – petri dishes would be suitable
- measuring jugs or cylinders
- test tubes or beakers in which to heat the fruit puree
- means of heating the fruit puree – water bath or Bunsen burner, tripod, gauze and heatproof mat
- access to a kettle and/or microwave oven
- access to a food processor to puree the fruit
- access to a fridge/freezer.

Chemicals

Each group of students will need:

- party jelly cubes – keep the pack to show students the warning that the jelly will not set if fresh pineapple, kiwi or papaya are added
- tinned pineapple
- fresh pineapple
- other fruits such as papaya, kiwi fruit, apple *etc*
- chillies if required for the extension activity.

Safety

- Wear eye protection.
- Make sure that students follow proper hygiene precautions (see the section ***Experiments with food*** in ***How to use this material*** on page vii).
- Take care with boiling water.
- Your employer's risk assessment should be consulted before carrying out this activity. This activity is covered by model (general) risk assessments widely adopted for use in UK schools and colleges such as those provided by CLEAPSS, SSERC, ASE and DfES. Bear in mind, however, that these may need some modification to suit local conditions.

Further information

The proteolytic enzyme found in pineapple is called bromelin, that found in papaya is called papain and that in kiwi fruit is actinidin.

Papain breaks protein chains at glycine molecules. It is used to tenderise meat by breaking down collagen and also to remove cloudiness from beer, which may also be caused by collagen.

Gelatine is extracted from meat and animal bones and is therefore not suitable for vegetarians. Carrageenan-based jellies are made from seaweed and are carbohydrate-based. They are not affected by papain and other proteolytic enzymes. These jellies are available in supermarkets and their response to various fruits could also be tested.

Both papain and bromelin are used commercially to 'tenderise' tough cuts of meat and some recipes call for the tenderising of meat by injecting it with pineapple juice.

Gels are not really solids or liquids. They are examples of **colloids**. These are materials in which one state of matter is dispersed within another – in the case of a gel, a liquid (water) is dispersed within a solid (the gelatine).

What effect does chilli have on the setting of jelly?

Teacher's notes:

....

....

....

....

....

....

....

....

....

....

....

KitchenChemistry
The chemistry of flavour
Star Kay White's
Pure
Chocolate Extract
Since 1890
A unique, fat free chocolate flavoring made by alcoholic extraction of a special blend of natural cocoa beans.
118 ml.
SUPER COOK
Natural
Lemon
Extract
SUPER COOK
Natural
Almond
Extract
Since
FINEST QUALITY
VANILLA EXTRACT
118 ml
Nielsen-Massey Vanillas,
Leeuwarden, NL

The chemistry of flavour

Learning objectives

- To devise and carry out experiments to test simple hypotheses.
- To develop the idea of control experiments.
- To develop awareness of analytical techniques such as gas chromatography and gas chromatography/mass spectrometry.
- To develop an understanding of the relationship between intermolecular forces and solubility.

Level

Age: post-16, but the experimental work could be used for any age group from 7 upwards.

Timing

The experimental work could be covered in a few minutes but could be developed into an open-ended investigation if desired. The reading and questions will take about 30 minutes and could be done for homework, as a revision exercise or in case of teacher absence.

Description

Some simple experiments are described that indicate that the sensation of flavour is made up of taste (detected by the tongue) and aroma (detected by the nose). A passage of written material about the chemistry of aroma-causing molecules is presented followed by questions which cover polarity, intermolecular forces and organic nomenclature.

Teaching notes

The lesson could be introduced by showing the video clips about detecting flavour and/or by carrying out an experiment similar to the 'blindfold' test (described in the activity below) as a demonstration with volunteers or as a class activity.

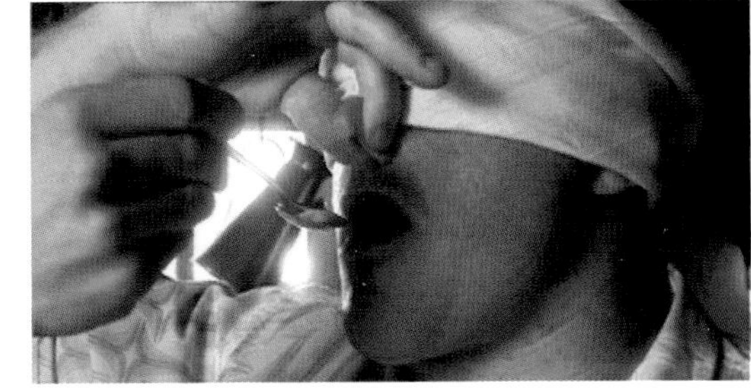

Index V04

Heston Blumenthal doing a blindfolded taste test

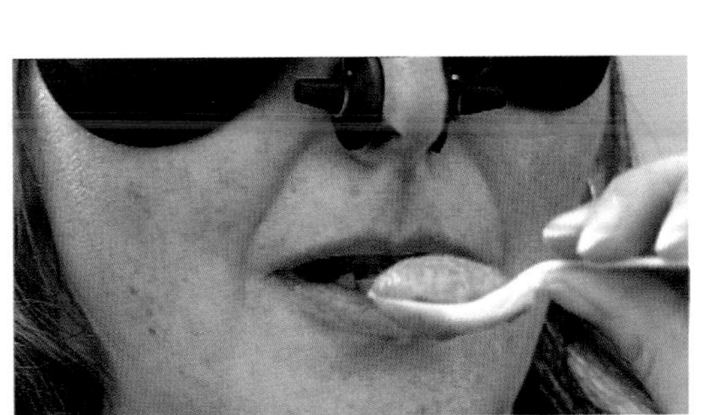

Index V05

A professional taste tester at work

Index V06

Heston Blumenthal (left) discussing the chemistry of taste with Peter Barham of Bristol University

Students may be surprised to find that detecting flavour requires both the tongue and the nose. We can think of flavour as being the sum of two sensations, taste and aroma. The tongue can detect five basic tastes. Four of these are well-known – sweetness (as in sugar), bitterness (as in quinine, found in tonic water, for example), saltiness and sourness (as in lemon juice). The fifth taste has only recently been recognised. It is called umami and is found in parmesan cheese, for example. The nose detects aromas. These are caused by small, volatile molecules that trigger receptors in the nose.

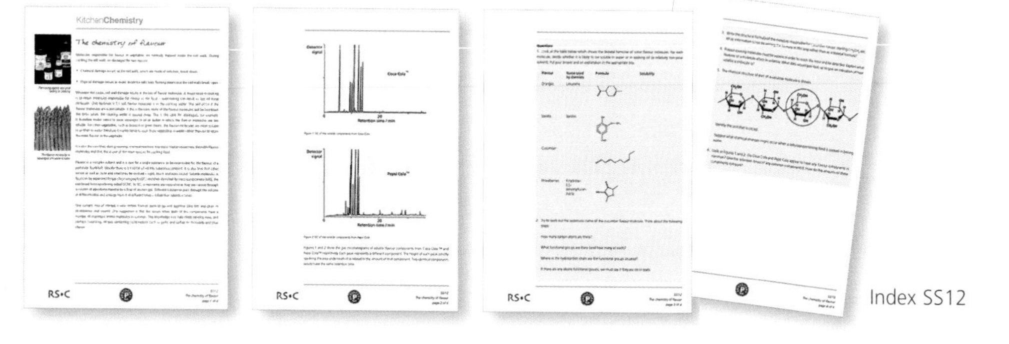

Index SS12

Activity

You can investigate the role of the nose in detecting flavour by setting up an experiment based on tasting foods with your nose blocked. A good example is strawberry jam. The taster must be blindfolded so that there are no clues from the appearance of the food and must hold his/her nose. S/he is fed a small amount of strawberry jam on a teaspoon. While holding his/her nose, s/he will be able to detect only the sweetness of the jam (which is detected using the tongue) but will be unable to tell the flavour of the jam which is caused by volatile molecules detected by the nose. On releasing his/her nose, s/he will be able to tell the flavour of the jam.

Other flavours of jam could be used as could a variety of foods such as apple, parsnip, broccoli *etc*, pureed so that there are no clues from the texture of the food.

A variation on this experiment uses potato crisps. Blindfold a volunteer and tell them that you are going to give them samples of crisps of different flavours to taste to see if they can identify the flavours. In fact, give them a plain crisp to taste and, without telling them, hold a flavoured crisp under their nose. They will usually believe that they are tasting a crisp of the flavour held under their nose, confirming that much of the sensation of flavour comes from the nose. The passage from the student's worksheet is reproduced below along with the answers to the questions.

The chemistry of flavour

Molecules responsible for flavour in vegetables are normally trapped inside the cell walls. During cooking the cell walls are damaged for two reasons:

- Chemical damage occurs as the cell walls, which are made of cellulose, break down.

- Physical damage occurs as water inside the cells boils forming steam, and the cell walls break open.

Whatever the cause, cell wall damage results in the loss of flavour molecules. A major issue in cooking is to retain molecules responsible for flavour in the food – overcooking can result in loss of these molecules. One destination for lost flavour molecules is in the cooking water. This will occur if the flavour molecules are water-soluble. If this is the case, many of the flavour molecules will be lost down the drain when the cooking water is poured away. This is the case for asparagus, for example. It therefore makes sense to cook asparagus in oil or butter in which the flavour molecules are less soluble. For other vegetables, such as broccoli or green beans, the flavour molecules are more soluble in oil than in water therefore it makes sense to cook these vegetables in water rather than oil to retain the most flavour in the vegetable.

It is also the case that, during cooking, chemical reactions may occur that produce new, desirable flavour molecules and that this is one of the main reasons for cooking food.

Flavour is a complex subject and it is rare for a single substance to be responsible for the flavour of a particular foodstuff. Usually there is a cocktail of volatile substances present. It is also true that other senses as well as taste and smell may be involved - sight, touch and even sound. Volatile molecules in food can be separated by gas chromatography (GC) and then identified by mass spectrometry (MS), the combined technique being called GCMS. In GC, components are separated as they are carried through a column of absorbent material by a flow of an inert gas. Different substances pass through the column at different rates and emerge from it at different times – called their retention times.

One current area of interest is why certain flavours seem to go well together (like fish and chips or strawberries and cream). One suggestion is that this occurs when both of the components have a number of important aroma molecules in common. This knowledge may help chefs develop new, and perhaps surprising, recipes containing combinations such as garlic and coffee or chocolate and blue cheese.

Index V07

Aparagus cooking in oil rather than water

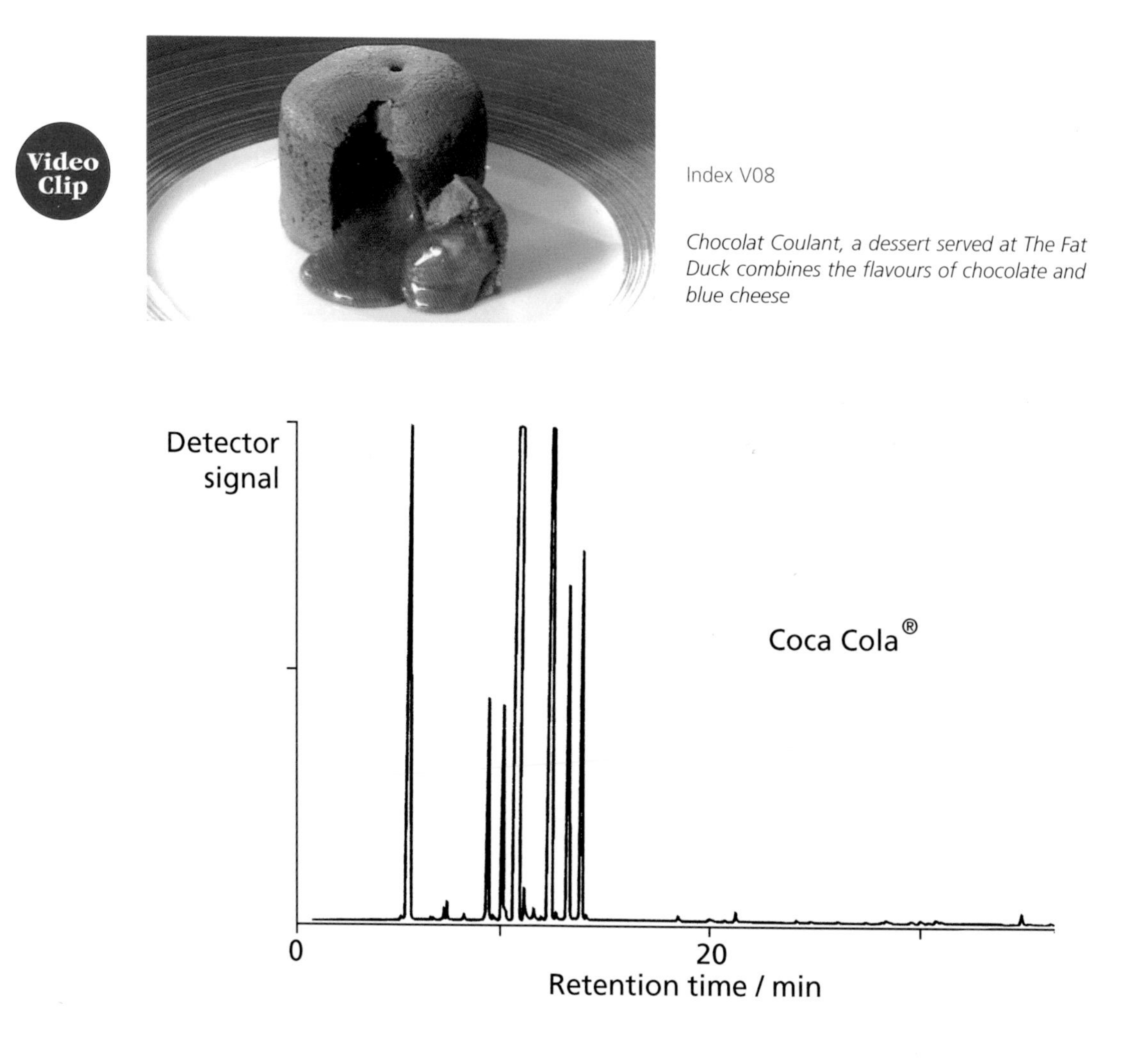

Index V08

Chocolat Coulant, a dessert served at The Fat Duck combines the flavours of chocolate and blue cheese

Figure 1 Gas chromatogram of the volatile components from Coca Cola®

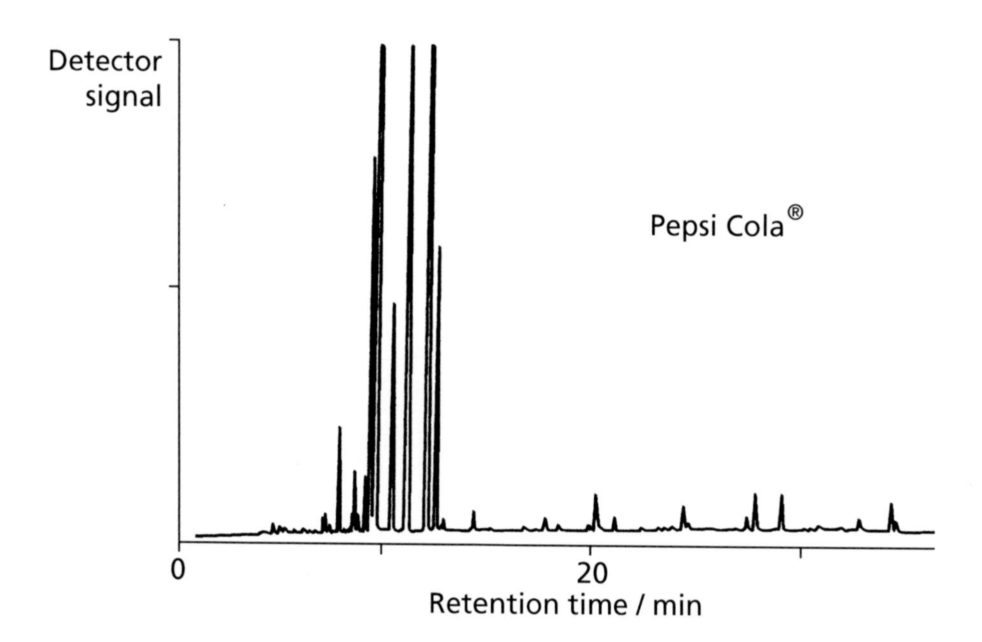

Figure 2 Gas chromatogram of the volatile components from Pepsi Cola®

Figures 1 and 2 show the gas chromatograms of volatile flavour components from Coca Cola® and Pepsi Cola® respectively. Each peak represents a different component. The height of each peak (strictly speaking the area underneath it) is related to the amount of that component present in the mixture. Two identical components would have the same retention time.

Questions

1. Look at the table below which shows the skeletal formulae of some flavour molecules. For each molecule, decide whether it is likely to be soluble in water or in cooking oil (a relatively non-polar solvent). Write down your answer and an explanation for each one.

Flavour	Name used by chemists	Formula
Oranges	Limonene	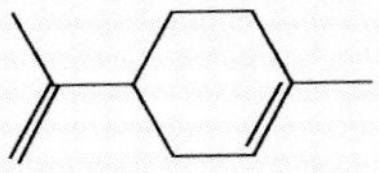
Vanilla	Vanillin	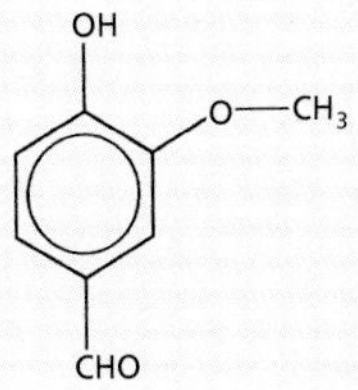
Cucumber	?	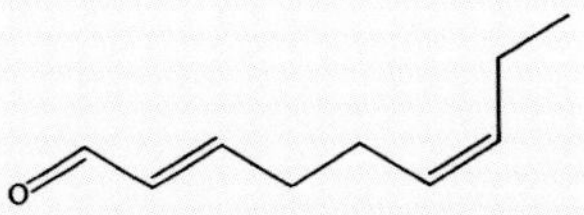
Strawberries	4-hydroxy-2,5-dimethylfuran-3-one	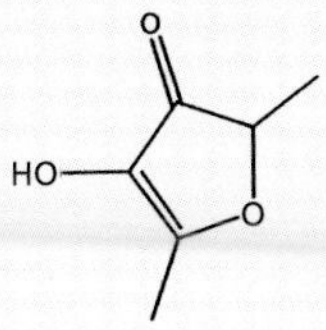

2. Try to work out the systematic name of the cucumber flavour molecule. Think about the following steps:

   How many carbon atoms are there?

   What functional groups are there (and how many of each)?

Where in the hydrocarbon chain are the functional groups situated?

If there are any alkene functional groups, we must say if they are *cis* or *trans*.

Write down your answer.

3. Write the structural formula of the molecule responsible for cucumber flavour, starting CH_3CH_2 *etc*. What information is lost by writing the formula in this way rather than as a skeletal formula?

4. Flavour-causing molecules must be volatile in order to reach the nose and be detected. Explain what features of a molecule affect its volatility. What data would you look up to give an indication of how volatile a molecule is?

5. The chemical structure of part of a cellulose molecule is shown below.

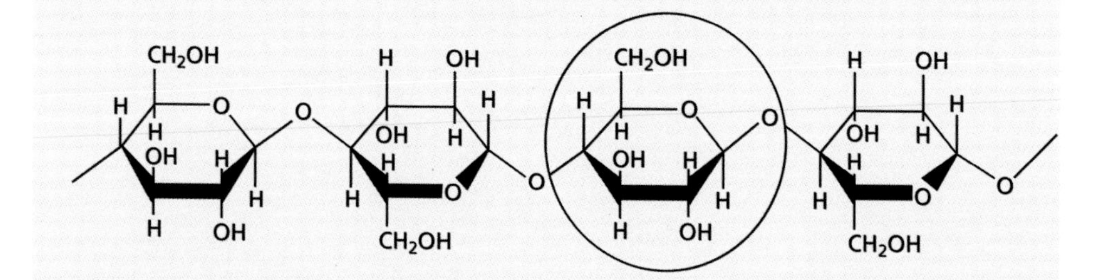

Identify the unit that is circled.

Suggest what chemical changes might occur when a cellulose-containing food is cooked in boiling water.

6. Look at Figures 1 and 2. Do Coca Cola® and Pepsi Cola® appear to have any flavour components in common? Give the retention times of any common component(s).

How do the amounts of these components compare?

Answers

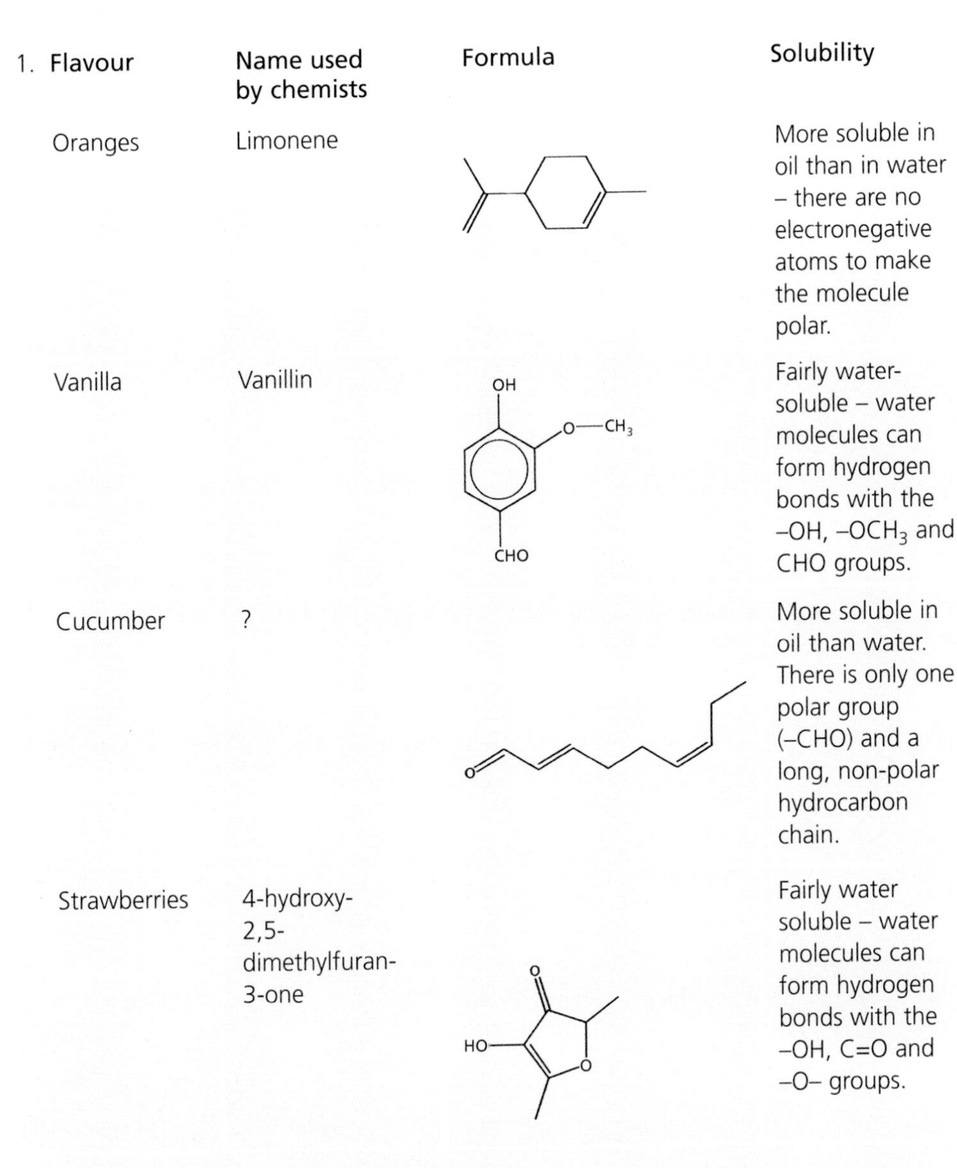

1. Flavour	Name used by chemists	Formula	Solubility
Oranges	Limonene		More soluble in oil than in water – there are no electronegative atoms to make the molecule polar.
Vanilla	Vanillin	OH, O—CH_3, CHO	Fairly water-soluble – water molecules can form hydrogen bonds with the –OH, –OCH_3 and CHO groups.
Cucumber	?	O	More soluble in oil than water. There is only one polar group (–CHO) and a long, non-polar hydrocarbon chain.
Strawberries	4-hydroxy-2,5-dimethylfuran-3-one	O, HO, O	Fairly water soluble – water molecules can form hydrogen bonds with the –OH, C=O and –O– groups.

2. 2-*trans*-6-*cis*-nonadienal

   There are nine carbon atoms, therefore the name is based on the root 'non'.

   There is an aldehyde (-al) functional group and two alkene groups (ene).

   The aldehyde is on the end of the chain (as it must be) and the alkenes are at carbon 2 and carbon 6 (counting from the aldehyde end).

   The alkene at carbon 2 is *trans* and that at carbon 6 is *cis*.

3. $CH_3CH_2CHCHCH_2CH_2CHCHCHO$

   The information lost is whether the two carbon-carbon double bonds are *cis* or *trans*.

4. Volatility is governed by intermolecular forces – the weaker these are, the more volatile the molecule. So molecules of low relative molecular mass will tend to be volatile (weak van der Waals forces). Non-polar molecules will tend to be volatile (weak or non- existent dipole-dipole forces). Molecules without N-H or O-H groups will tend to be volatile as they cannot form hydrogen bonds. The boiling point of a compound will give an indication of its volatility – the lower the boiling point, the more volatile the compound.

5. The circled unit is a sugar (glucose) molecule.

   The –O– links between the sugar molecules may hydrolyse (react with water) to give shorter chains of sugar molecules.

6. The peak on the GC for Pepsi Cola® (Figure 2) at retention time approximately 12 minutes 20 seconds appears to be common to both drinks. The amounts of this component are similar because the peak height is approximately the same in both cases.

Teacher's notes:

Kitchen**Chemistry**

Chemical changes during cooking

Learning objective

- Revision of some organic reactions.

Level

Age: Post-16.

Timing

About 30 minutes.

Description

A passage of reading material is presented that describes some of the chemical reactions that take place on cooking food, in particular three types of reaction that cause food to brown. Questions are included that test familiar chemistry in an unfamiliar context. The topic covered is organic reactions. The exercise could be used for revision, homework or in the case of teacher absence.

Teaching notes

The lesson could be started by introducing Heston Blumenthal as a 'scientific chef' and showing the video clip of him discussing the cooking of meat.

Index V09

Temperature probes used when cooking meat at The Fat duck

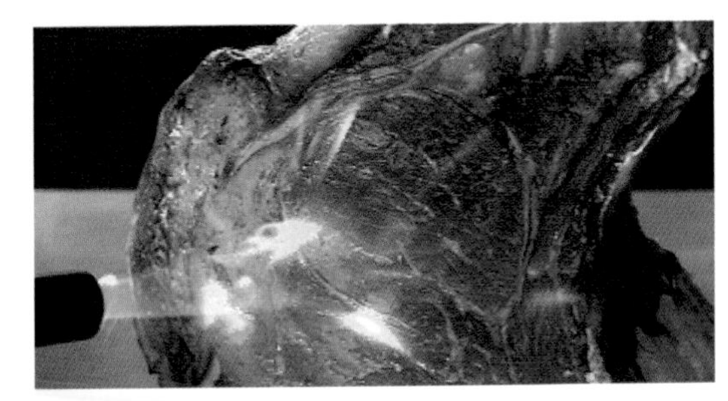

Index V10

A blowtorch is used to brown the surface of raw meat

If required, practical work to investigate the browning of cut fruit is available that could be used to develop the theme of browning reactions– see *Further information* on page 94. The passage in the student's sheet is reproduced overleaf along with the answers to the questions.

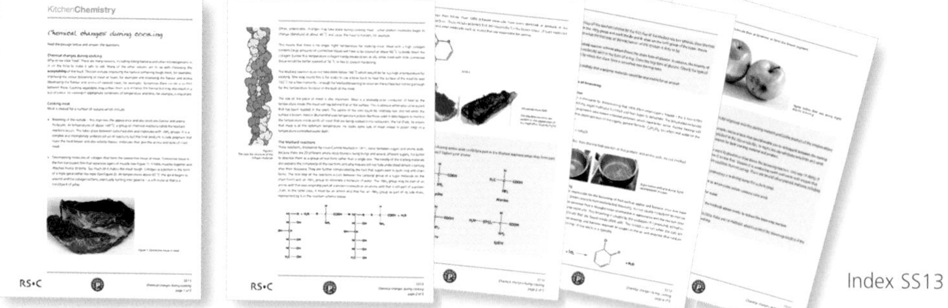

Index SS13

Chemical changes during cooking

Why do we cook food? There are many reasons, including killing bacteria and other microorganisms in or on the food to make it safe to eat. Many of the other reasons are to do with increasing the **acceptability** of the food. This can include improving the texture (softening tough meat, for example), improving the colour (browning of meat or toast, for example) and improving the flavour and aroma (developing the flavour and aroma of cooked meat, for example). Sometimes there can be a conflict between these goals. Cooking vegetables may soften them and enhance the flavour but may also result in a loss of colour. So using appropriate cooking conditions, such as temperature and cooking time, is important.

Cooking meat

Meat is cooked for a number of reasons which include:

- Browning of the outside – this improves the appearance and also produces flavour and aroma molecules. At temperatures of above 140 °C a group of chemical reactions called the Maillard reactions occurs. These take place between carbohydrates and molecules with $-NH_2$ groups. They are a complex and incompletely understood set of reactions but the final products include polymers that make the meat brown and also volatile flavour molecules that give the aroma and taste of roast meat.

- Decomposing molecules of collagen that form the connective tissue of meat. Connective tissue is the thin translucent film that separates layers of muscle (see Figure 1). It holds muscles together and attaches muscle to bone. Too much of it makes the meat tough. Collagen is a protein in the form of a triple spiral rather like rope (See Figure 2). At temperatures above 60 °C the spiral begins to unwind and the collagen softens, eventually turning into gelatine – a soft material that is a constituent of jellies.

Figure 1 Connective tissue in meat

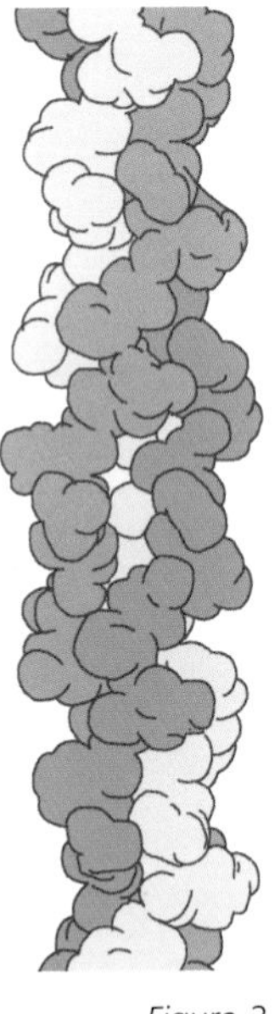

Figure 2
The rope-like structure of the collagen molecule

Other, undesirable changes may take place during the cooking of meat - other protein molecules begin to change (denature) at about 40 °C and cause the meat to harden, for example.

This means that there is no single 'right' temperature for cooking meat. Meat with a high collagen content (large amounts of connective tissue) will have to be cooked at above 60 °C for some time to break down the collagen (below this temperature collagen hardly breaks down at all), while meat with little connective tissue would be better cooked at a slightly lower temperature to prevent the proteins denaturing too much and forcing out precious juices.

The Maillard reactions do not take place below 140 °C, which would be far too high a temperature for cooking. One way round this is for cooks to use a blow torch to heat the surface of the meat to over 140 °C for a few moments – enough for Maillard browning to occur on the surface but not long enough for this temperature to occur in the bulk of the meat.

The size of the piece of meat is also important. Meat is a relatively poor conductor of heat so the temperature inside the meat will lag behind that of the surface. This is obvious when you carve a joint that has been roasted in the oven. The centre of the joint could be relatively raw and red while the surface is brown. Heston Blumenthal uses temperature probes like those used in data loggers to monitor the temperature inside joints of meat that are being cooked in his restaurant, The Fat Duck, to ensure that meat is at the optimum temperature. He cooks some cuts of meat sealed in plastic bags in a temperature-controlled water bath.

The Maillard reactions

These reactions, discovered by Louis-Camille Maillard in 1911, occur between sugars and amino acids. Because there are 20 different amino acids found in living things and several different sugars, it is better to describe them as a group of reactions rather than a single one. The variety of the starting materials also explains the complexity of the reactions and why they are still not fully understood almost a century after their discovery. They are further complicated by the fact that sugars exist in both ring and chain forms. The first step of the reactions occurs between the carbonyl group of a sugar molecule (in the chain form) and an $-NH_2$ group, to eliminate a molecule of water. The $-NH_2$ group may be part of a free amino acid molecule (that was originally part of a protein molecule) or an amino acid that is still part of a protein chain. In the latter case, it must be an amino acid that has an $-NH_2$ group as part of its side chain. This side chain is different for each amino acid and is represented by $R-NH_2$ in the reaction scheme below.

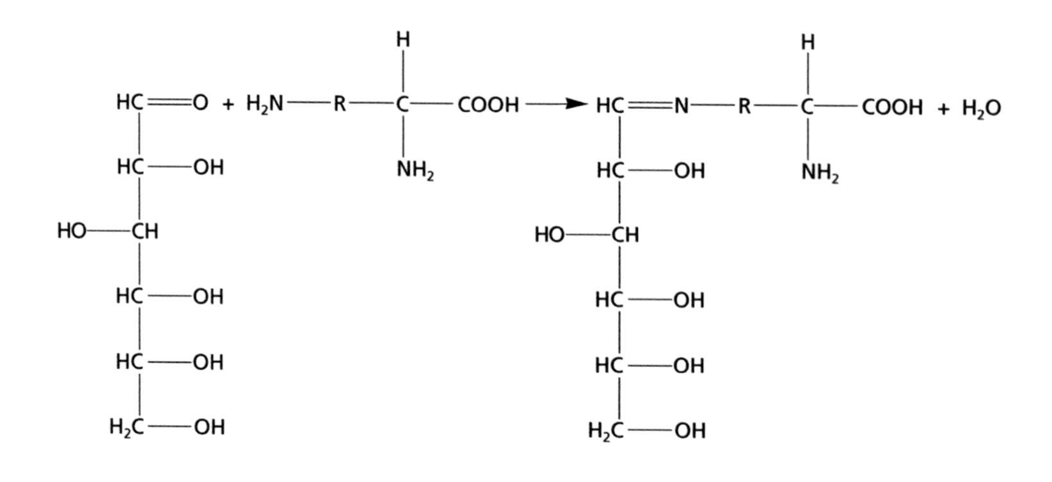

Further steps then follow. Over 1000 different molecules have been identified as products of the Maillard reactions. These include polymers that are responsible for the brown colour of roast meat (and toast, *etc*) and small molecules such as maltol that are responsible for aromas.

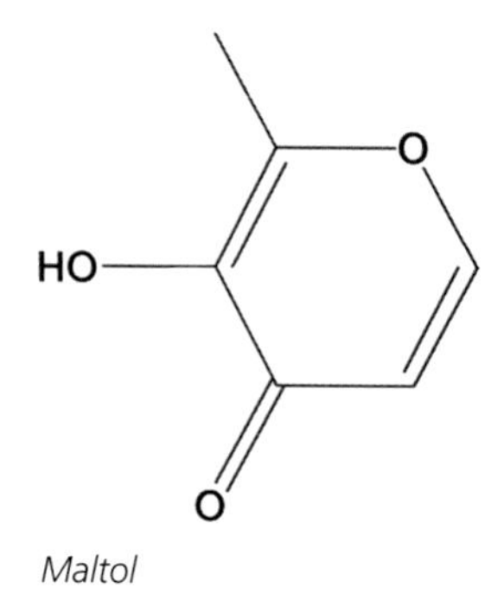

Maltol

Questions

1. Which of the following amino acids could take part in the Maillard reactions when they form part of a protein chain? Explain your answer

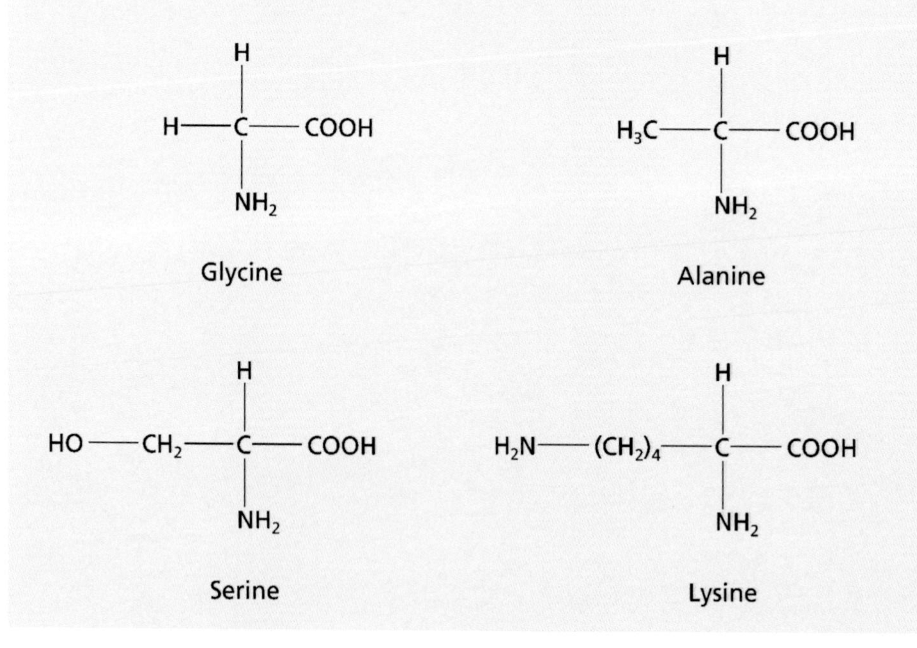

2. On a copy of the reaction scheme for the first step of the Maillard reaction, draw the lone pair on the $-NH_2$ group and mark the δ+ and δ- areas on the C=O group of the sugar. Hence suggest what the first step of the mechanism of this reaction is likely to be.

3. The Maillard reaction scheme above shows the chain form of glucose. In solution, the majority of glucose molecules exist in the form of a ring. Draw the ring form of glucose. Classify the type of reaction by which the chain form is converted into the ring form.

4. Why is it unlikely that a polymer molecule would be responsible for an aroma?

Other types of browning

Caramelisation

Caramelisation is the name for the browning that takes place when sugar is heated – this is how toffee is made. On heating, sugar molecules first melt and then begin to dehydrate. The dehydrated molecules then begin to polymerise into brown-coloured polymers which have bitter tastes. Further heating will result in complete decomposition of the sugars, general formula $C_n(H_2O)_n$, to carbon and water (in the form of steam):

$$C_n(H_2O)_n \rightarrow nC + n(H_2O)$$

This reaction differs from the Maillard reaction in that proteins and amino acids are not involved.

Sugar before (left) and during (right) caramelisation in a pan

Apples before (left) and during (right) enzymic browning

Enzymic browning

Enzymic browning is responsible for the browning of fruit such as apples and bananas once they have been cut or bruised. Unlike caramelisation and Maillard browning, it is not usually considered to improve the food because the browned fruit is thought to be unattractive in appearance and the reaction does not result in any aroma molecules. This browning is caused by the oxidation of compounds related to phenols (hydroxybenzenes) that are found inside plant cells. The oxidation occurs when the cells are damaged by cutting or bruising and become exposed to oxygen in the air and enzymes that catalyse the oxidation. The first step of the reaction is typically:

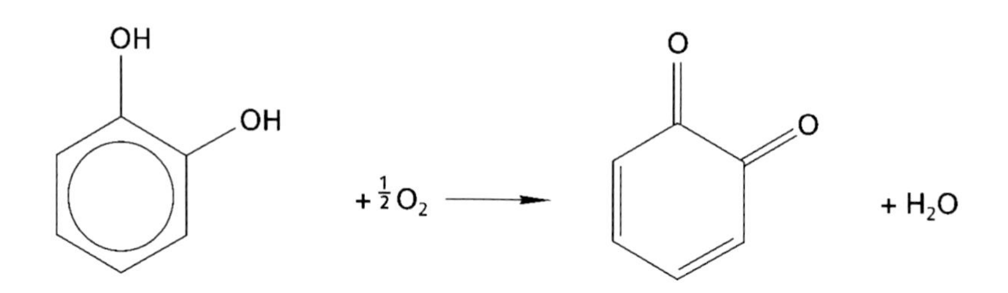

The product molecule then polymerises to form the brown pigment.

Questions

5. (a) Name the functional groups in (i) the starting material and (ii) the product of the reaction above.

   (b) Suggest two simple chemical tests that would enable you to distinguish between the starting material and the product in the above reaction. In each case say what you would do and what result you would expect for both starting material and product.

6. (a) Cooks normally want to prevent or slow down the browning reaction. One way of doing this is to add vitamin C to the fruit. Vitamin C is a reducing agent and reacts with oxygen thus preventing oxidation reactions from occurring. There are several other possible methods including:

   Blanching the fruit (immersing it in boiling water for a short time).

   Adding an acid (such as lemon juice, which contains citric acid).

   Storing the fruit under water.

   Suggest how each of the methods above works to reduce the browning reaction.

   (b) Suggest what restrictions there are on methods used to control the browning reaction if the fruit is to be used in cooking.

Further information

Experimental work to investigate the browning of fruit is described in a number of sources including:

Nuffield Food Science, a Special Study, London: Longman, 1974

Revised Nuffield Food Science, a Special Study, London: Longman, 1984

Note. Information about the Nuffield Chemistry Special Study materials is available at *www.chemistry-react.org (accessed January 2005)*

The Most Useful Science, Harleston: Thorburn Kirkpatrick, 1989

Answers

1. Lysine – it has an $-NH_2$ group on its side chain which will be free to take part in a Maillard reaction even when its other amino group is involved as part of a peptide link.

2. The lone pair on the $-NH_2$ group can be shown as NH_2. The C=O group is polarised $C^{\delta+}=O^{\delta-}$. The first step is likely to be a nucleophilic attack in which the lone pair of the $-NH_2$ group attacks and forms a bond with the $C^{\delta+}$.

3.

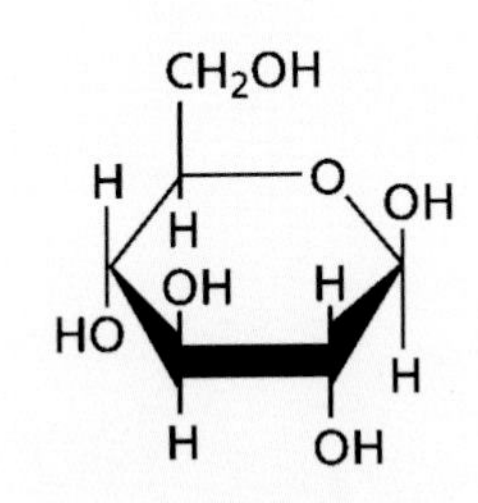

This is a nucleophilic addition reaction (in which a lone pair on one of the –OH groups attacks the carbon atom on the –C=O group).

4. Aroma molecules are detected by the nose and must therefore be volatile. Polymer molecules are too large to be volatile.

5. (a) (i) Phenols
   (ii) Ketones

   (b) There are several possibilities:

   The starting material will react with sodium to give off hydrogen while the product will not. Warm a little of the organic material in a test tube until it melts and add a small cube of sodium – the starting material will fizz and the product will not.

The starting material is acidic and the product is not.

A solution of the starting material will give a purple colouration when iron(III) chloride solution is added, the product will not.

The product will form an orange precipitate of a 2,4-dinitrophenylhydrazone with Brady's reagent (2,4-dinitrophenylhydrazine) while the starting material will not. Add a little of the organic material to a little 2,4-dinitrophenylhydrazine dissolved in acidified methanol. The starting material will not react but the product will form an orange precipitate.

6. (a) Blanching will denature the enzyme catalyst.

   Adding acid will denature the enzyme and/or change the pH to take it outside the optimum range for the enzyme.

   Storing the fruit under water will help prevent oxygen getting to the fruit.

   (b) Any treatment must leave the fruit fit for human consumption, *ie* safe to eat and with no unacceptable flavours, aromas or colours.

Teacher's notes:

KitchenChemistry

The science of ice cream

Learning objectives

- Understanding the relationship between intermolecular forces and solubility.
- Knowledge of some of the factors affecting crystallisation.

Level

Making ice cream using liquid nitrogen as a coolant is suitable for any age group and would be a good demonstration for an end of term lesson or open day. The level of discussion about the chemistry of the process could be tailored by the teacher to any age from primary to post-16.

Timing

About 30 minutes allowing time for discussion (and tasting!) as well as the making of the ice cream. The world record for making a litre of ice cream using liquid nitrogen is 20.91 seconds at the time of writing.

Description

A recipe is given for making ice cream using liquid nitrogen as the cooling agent. This could be carried out along with discussion of the need for rapid cooling and/or stirring to produce smooth ice cream with small ice crystals. Some discussion of issues related to the flavour of ice cream is also presented.

Teaching notes

The lesson could be started by introducing Heston Blumenthal as a 'scientific chef' and showing the video clip of him discussing the making of ice cream.

Index V11

A dish of red cabbage gazpacho with grain mustard ice cream served at The Fat Duck

Ice cream consists mainly of:

- ice
- fat – from milk or cream
- sugar
- air.

Sometimes, egg yolk and other ingredients are added to give it flavour and colour.

Most people prefer smooth ice cream with no lumps of ice, which would give it a gritty feel in the mouth. An understanding of the chemistry of ice cream helps to see how this smoothness is achieved.

To make ice cream conventionally, mix egg yolks and milk and stir in sugar along with

flavouring such as vanilla. Stirring at this stage helps to prevent lactose molecules forming crystals which could give the ice cream a gritty texture. (Lactose is the main sugar found in milk.) Then heat the mixture, while stirring, but to no more than 65 °C. Above this temperature, the protein molecules in the egg begin to denature and coagulate into lumps – effectively the same process that occurs when making scrambled egg. The product at this stage is in effect a custard and has to be frozen to make it into ice cream. This was traditionally done by cooling the custard in a container in 'freezing mixture' (a mixture of ice and water whose temperature can be as low as -20 °C). During the cooling process, stirring is essential. This is because ice crystals form during cooling. If they are not broken up by stirring, these crystals can grow large and give the ice cream a gritty texture.
Note; an animated explanation of how adding salt lowers the freezing point of ice can be viewed at ***http://antoine.frostburg.edu/chem/senese/101/solutions/faq/why-salt-melts-ice.shtml*** *(accessed January 2005)*

A general rule about growing crystals is that if they form slowly, we get a small number of large crystals and if they form quickly we get large numbers of small crystals. Contrast the small crystals obtained by boiling the water from a solution of copper sulfate, say, with the large ones obtained by allowing the water to evaporate over several days in a warm place. Over a vastly larger time scale, when magma cools slowly it forms granite rock with large crystals while more rapid cooling forms basalt, which has smaller crystals.

One way to make ice cream crystals quickly is to cool the custard by adding liquid nitrogen. This liquid is at -196 °C and is constantly in the process of boiling. If liquid nitrogen is poured into an ice cream custard, it cools within seconds (the world record is about 20 seconds for a litre (1 dm^3) of ice cream) and a smooth ice cream with very tiny ice crystals is formed. The nitrogen boils away forming harmless nitrogen gas and the ice cream is quite safe to eat once it has warmed up to about 0 °C.

Making ice cream using liquid nitrogen

Video Clip

Index V13

A world record attempt:
ice cream making using liquid nitrogen

Liquid nitrogen is used in university research laboratories, hospitals and other facilities. It is quite cheap – a few pence per litre – and many institutions are prepared to give away a litre or two for educational purposes. A purpose-built Dewar (vacuum flask) is required for transport but the donating institution may well be prepared to loan one. Making ice cream with liquid nitrogen is quite spectacular and easy, providing liquid nitrogen is available and suitable precautions are taken for transporting and working with it – see the risk assessments in the Appendix on page 103.

Heston Blumenthal using liquid nitrogen in his kitchen laboratory

Ingredients

- 4 egg yolks
- 500 cm^3 milk
- 500 cm^3 cream (single or double)
- 120 g sugar
- 300 g fresh strawberries, pureed (other fresh or tinned fruit can be substituted)
- 1 dm^3 liquid nitrogen.

Method

Follow strict hygiene precautions, especially if working in a laboratory (see the section ***Experiments with food*** in ***How to use this material*** on page vii). Mix the ingredients in a large metal bowl (this avoids the possibility of cracking a ceramic or glass bowl by thermal shock) standing on a cork mat (to provide thermal insulation) and stir well. The bowl should be large enough to hold four times the volume of the ice cream to prevent splashing when the liquid nitrogen is added. A long-handled wooden spoon gives good thermal insulation. Wearing insulating gloves and eye protection, pour in about half the liquid nitrogen. This will boil vigorously and produce large quantities of 'fog' (water vapour condensed from the air). Stir again. If the mixture has not all frozen, add more of the liquid nitrogen. After a few moments, check that all the liquid nitrogen has evaporated. As the ice cream begins to soften it can be tasted. In view of the unfamiliar and spectacular properties of liquid nitrogen, some students may need reassurance before tasting the ice cream.

Safety

- Wear eye protection, ideally a face shield, when using liquid nitrogen.
- Beware of 'cold burns' from liquid nitrogen – wear insulating gloves.
- Make sure that students follow proper hygiene precautions when tasting.
- Use disposable plastic spoons (one per student) for tasting.

- Your employer's risk assessment should be consulted before carrying out this activity. Special risk assessments have been produced by CLEAPSS for the transport and use of liquid nitrogen. See the Appendix on page 103.

Students tasting ice cream

The flavour of ice cream

One of Heston Blumenthal's recipes is for vanilla, chocolate and pistachio flavoured ice cream. This ice cream contains three flavour ingredients – vanilla, chocolate and pistachio paste. On tasting the ice cream, the vanilla flavour is noticed first followed some time later by the chocolate and pistachio. This effect occurs because the molecule responsible for vanilla flavour – vanillin – is relatively soluble in water. The flavours of chocolate and of pistachio are caused by mixtures, rather than a single compound and the substances involved are fat-soluble.

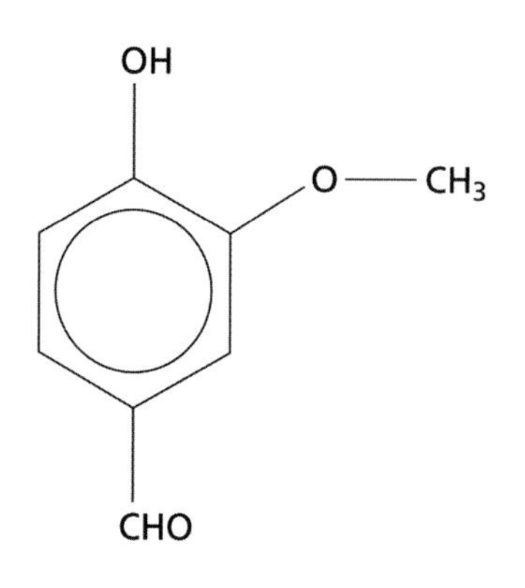

Vanillin

As the ice cream melts in the mouth, the ice (in which the water-soluble vanillin is trapped) melts rapidly and releases the vanilla flavour. The fats (which hold the pistachio and chocolate flavours) melt at a higher temperature and do not release their flavours for a little longer. Heston Blumenthal discusses this in the video clip.

Index V12

Vanilla, chocolate and pistachio ice cream

Question (post-16)

Water-solubility arises when a molecule can form intermolecular bonds with water molecules. What sort of intermolecular bond can vanillin form with water molecules? What features of the vanillin molecule mean that it can form this type of bond? Draw a diagram to show one of these bonds forming.

Answer

Hydrogen bonding. Hydrogen bonds will form between a hydrogen atom on a water molecule and any of the oxygen atoms in the vanillin molecule. A hydrogen bond will also form between the hydrogen of the –OH group of the vanillin and an oxygen atom of a water molecule.

Appendix – risk assessments

CLEAPSS® School Science Service

Risk Assessment (to meet the COSHH or Management Regulations)

Applicant:	Education Manager (Schools & Colleges)
School / LEA:	Royal Society of Chemistry
Operation:	Transport and use of liquid nitrogen for teacher demonstrations

Details of operation:

A variety of simple demonstrations are to be performed using liquid nitrogen.

(a) Typical activities include the following: freezing flowers, rubber tubing, banana, inflated balloon; whistling kettle, inflating balloons by attaching to corked Buchner flask.

(b) A significantly different activity involves making (and subsequently tasting) ice cream, by mixing liquid nitrogen with egg yolks, milk, sugar, etc.

It is the responsibility of the applicant to inform the CLEAPSS® School Science Service if these details of the operation are substantially inaccurate.

Note:

(1) **This risk assessment assumes that the activities are to be carried out by a qualified science teacher. If pupils/students are to handle liquid nitrogen a further risk assessment would be necessary.**

(2) **In addition to its use in demonstrations as above, this risk assessment also considers issues about the transport and handling of liquid nitrogen.**

Substance(s) possibly hazardous to health:	Liquid nitrogen
Classification under CHIP2 Regulations 2002	Not listed
Particular risks/precautions:	1 Asphyxiation in oxygen deficient atmospheres 2 Fire in oxygen enriched atmospheres 3 Cold burns, frost bite & hypothermia from the intense cold 4 Over-pressurisation from the large volume expansion of the liquid 5 Manual handling accidents if using large (25 litre) volumes.
Maximum exposure limits:	Not listed
Occupational exposure standards:	Not listed

Risk assessment

General points

1 **Liquid nitrogen should be handled only by teachers and technicians, not by pupils. Teachers must practice demonstrations before showing them to an audience.**

2 If the brittleness of frozen rubber and similar activities are to be demonstrated, beware the risk of sharp splinters and very cold fragments causing injury. Unless the demonstrator is well separated from the audience, observers should wear eye protection. Safety screens may also be required.

Asphyxiation in oxygen deficient atmospheres

3 Normal dry air contains 20.95% oxygen by volume. Industrial practice is designed to ensure that oxygen in the workplace never falls below 18% and entry into atmospheres less than 19.5% is not recommended. Asphyxiation due to oxygen deficiency can be rapid with no prior warning. Liquid nitrogen when it evaporates to form the gas will be cold and hence denser than normal air. A layer of nitrogen will tend to accumulate at floor level until it diffuses away. The death of a worker in a university was reported some years ago where liquid nitrogen evaporated into a cellar area.

4 If 2 litres liquid nitrogen all evaporated at once it would give about 1.37 m^3 nitrogen gas (at room temperature and pressure). If well mixed, this could be released into a room of capacity 91 m^3 without falling below the 19.5% figure. Most laboratories are much larger than this, typically 240 m^3 but prep rooms are likely to be smaller.

5 **Small amounts (not exceeding 2 litres) of liquid nitrogen can be safely used in school laboratories, without risk of asphyxiation. (For larger amounts, see later). It would be prudent to ensure good ventilation. However, if it is kept in preparation rooms, stores and other confined spaces these must be ventilated as much as possible. In the event of a spill; the room should be evacuated. If liquid nitrogen has been kept in a confined space, when retrieving it a second person should be present in case of asphyxia.**

Fire in oxygen enriched atmospheres

6 The boiling point of nitrogen is -196 °C. The boiling point of oxygen is -183 °C. Thus liquid nitrogen may condense liquid oxygen from the surrounding air. Substances will burn more fiercely and ignite more easily in oxygen-enriched atmospheres than in normal air.

7 **Provided that the quantities of liquid nitrogen are small (eg, 2 litres) we do not consider that there is much risk of significant oxygen enrichment.**

Cold burns, frost bite & hypothermia from the intense cold

8 There is a temperature difference of 233 °C between liquid nitrogen and the human body. Contact with skin will result in burns in much the same way as contact with something 233 °C hotter than the human body would do. Contact with cold vapour or cold objects will have a similar effect. If small amounts of liquid nitrogen are splashed on the skin, the warmth of the body instantly vaporises a tiny amount of liquid nitrogen. This may act as an insulating layer preventing contact between skin and the bulk of the nitrogen, thus sometimes preventing burns. It is, however, unwise to rely on this as a preventative measure.

9 Liquid nitrogen can become entrapped in clothing, including shoes and gloves.

10 **When handling liquid nitrogen, eye protection (preferably goggles or a face shield) must be worn and non-absorbent leather gloves. Shoes and gloves need to be easily removable but open-toed shoes should NOT be worn. Demonstrators need to be vigilant about the possibility of splashes of liquid nitrogen becoming entrapped in**

clothing, etc.

Over-pressurisation from the large volume expansion of the liquid

11 When liquid nitrogen at -196 °C vaporises to the gas at room temperature, the volume increases by a factor of approximately 683. Thus 2 litres become 1366 litres. In an enclosed vessel this will give rise to a huge increase in pressure. Because liquid nitrogen is cold, it will condense any moisture in the atmosphere possibly forming an ice plug which may seal an open vessel, causing the pressure to build up. To our knowledge this has caused at least two explosions in schools/colleges, although fortunately no injuries.

12 **Liquid nitrogen must be transported, kept and used only in a vented Dewar flask, specially designed for cryogenic work. Do NOT, under any circumstances, use ordinary vacuum flasks.** (Suitable small Dewars are available for less than £100, eg, from LabSales (01954 233190).

Consumption of ice cream made with liquid nitrogen

13 If it is intended to use liquid nitrogen to make ice cream, then there are additional hazards if the ice cream is to be consumed.

14 Most of the liquid nitrogen produced by companies such as BOC is used for fast freezing food, eg, burgers. There is therefore nothing present which would render it unsuitable for human consumption.

15 **Liquid nitrogen from most sources should be safe for human consumption** (but see below).

16 It is an offence under the COSHH Regulations to permit eating, drinking, etc in any area which could be contaminated by hazardous chemicals or microorganisms. This would include most school laboratories, unless scrupulous methods have been adopted to ensure no possibility of contamination. This would apply even more to any equipment used to prepare ice cream. Food technology rooms and equipment, or school canteens are preferable.

17 **Preparation of ice cream using liquid nitrogen should not normally take place in laboratories. Similarly, normal science equipment should not be used – equipment should be kept in clean conditions and reserved for this task.**

18 Liquid nitrogen is very cold (see 6, 7, 8 above). If ice cream is consumed too soon, it will cause cold burns.

19 **Demonstrators must check the temperature of the ice cream before permitting its consumption. The thermometer must be scrupulously clean.**

20 There is a risk of food poisoning from the dairy products and eggs. Young children (age less than 2 years), the elderly, pregnant women, those taking drugs which suppress the body's natural ability to fight infection and people with reduced immunity (including AIDS sufferers, alcoholics, drug abusers, diabetics, transplant patients, those who are already ill or convalescent and those with known allergies) are at particular risk. Raw eggs present particular risks of salmonella. Some ice cream is made without eggs and this may be a preferred alternative.

21 **The food components in the ice cream must be fit for human consumption. Sugar must be specially purchased for this activity – do not use ordinary laboratory stocks. Milk and cream should be fresh. If eggs are to be used, they should come from flocks which are *salmonella* free. Those stamped with a lion mark are claimed to be. At risk groups should not consume ice cream made with raw eggs and the**

demonstrator should warn the audience.

Transport of liquid nitrogen

22 Under the Transportable Pressure Vessels Regulations 2001, liquid nitrogen must only be transported in vessels which are suitable for this purpose. Open vessels are NOT suitable, because of the risk of splashing and spills. A properly designed cryogenic vessel must be used. A vented Dewar must be used, a normal vacuum flask is NOT suitable. Accidents have occurred when using ordinary vacuum flasks. Suitable small Dewars, 1 or 2 litre capacity, are available for less than £100, eg, from LabSales (01954 233190).

23 If liquid nitrogen is transported by road, the Carriage of Dangerous Goods and Use of Transportable Pressure Equipment Regulations require that the driver shall have had special training. The BCGA Code of Practice CP30 The Safe Use of Liquid Nitrogen Dewars up to 50 litres (British Compressed Gases Association, 2000) states that Dewars must be transported separately from driver or passengers. Flat-back vehicles, trailers or vehicles fitted with a separating bulkhead should be used. These two requirements mean it is unlikely that schools will be able to obtain free supplies from a friendly source unless within walking distance. Hospitals, food processing companies, universities and even some doctor's surgeries use liquid nitrogen, so this may be possible.

24 If, exceptionally, small amounts of liquid nitrogen are transported by road, it is important to remember that small capacity Dewars have a relatively low base area. Thus there is a distinct risk of the Dewar falling over, eg, if the driver has to brake suddenly. This could lead to a major spill of liquid nitrogen, increasing greatly the risk of asphyxiation. Thus the Dewar must be transported in such a way that it cannot fall over or spill, for example by standing it in a large, deep cardboard or plastic box filled with crumpled newspaper or similar packing material.

Particular risks when using large volumes of liquid nitrogen

25 Rather than acquiring a small amount of liquid nitrogen from a local user, a school can pay to have it delivered, but in that case it is likely to be in 25 litre Dewars. A special stand will be needed, and can usually be hired at the same time as the Dewar.

26 Because of the bulk and weight of a 25 litre Dewar, a risk assessment will be required under the Manual Handling Regulations. Trolleys should be used to move the Dewar around. A tipping stand can be used to dispense the liquid nitrogen into smaller containers but two people will be required to carry the Dewar and mount it in the stand. If lifts have to be used, the Dewar should not be accompanied in the lift.

27 The larger volume of liquid nitrogen increases the asphyxiation risk. Slow evaporation in a store room or other confined space could cause problems. Ensure good ventilation. When retrieving it a second person should be at the door in case of asphyxia.

First aid for minor exposures to liquid nitrogen

The aim of treatment is to raise the temperature of the affected part slowly back to normal.

1. Move the victim to a warm room (22°C) if possible.
2. Loosen any restrictive clothing. Do not remove clothing that is stuck to the body until thawed thoroughly.
3. Place the affected area in TEPID WATER or flush the affected area with copious quantities of TEPID WATER for half an hour until the skin changes from pale yellow through blue to pink or red. Do not use hot water or any other form of direct heat such

as room heaters.

4. Keep the patient warm and at rest.
5. Cover the affected part with a bulky, dry, sterile dressing. Do NOT apply too tightly so that it restricts blood circulation.
6. Send the victim to a hospital casualty department. Ensure the ambulance crew is advised of the nature of the accident and the treatment provided so far.

References

1. BCGA Code of Practice CP30, The Safe Use of Liquid Nitrogen Dewars up to 50 litres, Eastleigh, British Compressed Gases Association, 2000.
2. SSERC Bulletin 190 Spring 1997 pp 6-8.

Assessor

Dr Peter Borrows, MA, PhD, CChem, FRSC

If further clarification is required, contact:

The CLEAPSS®School Science Service, *at* Brunel University, Uxbridge UB8 3PH
Tel: 01895 251496 *Fax / Answerphone:* 01895 814372
E-mail: science@cleapss.org.uk *Web site:* www.cleapss.org.uk

Teacher's notes:

Kitchen**Chemistry**

'Asparagus pee'

'Asparagus pee'

Learning objective

- Revision of amino acids, mass spectrometry, organic synthesis, optical activity.

Level

Age: Post-16.

Timing

About 20 minutes.

Description

The main part of this material is based on a question taken from the paper used by the RSC to select the team to represent the UK in the XXXVIth International Chemistry Olympiad, 2004. It is presented here because of its relevance to the chemistry of food. Topics covered include amino acids, mass spectrometry, synthetic organic chemistry and optical activity. The topic could be introduced using the information below.

The section on ***The chemistry of flavour*** on page 77 refers to the desirability of cooking asparagus in butter/oil rather than in water. This is because the water-solubility of the molecules responsible for the flavour means that much of the flavour will be literally poured down the drain if asparagus is cooked in water. Teachers might like to refer to this and also to the related video clip.

Video Clip

Index V07

Aparagus cooking in oil rather than water

Teachers might also like to ask students to consider the likely solubility of some of the compounds in the reaction scheme in question 2.

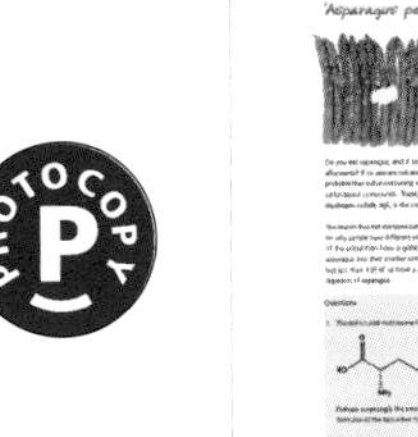

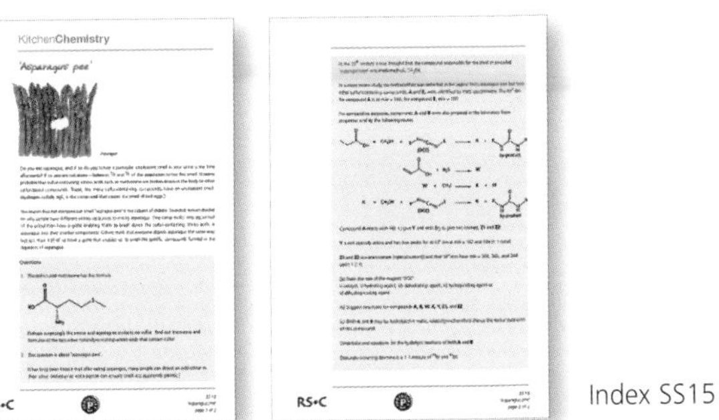

Index SS15

Teaching notes

'Asparagus-pee'

Do you eat asparagus, and if so do you notice a particular unpleasant smell in your urine some time afterwards? If so you are not alone – between $^1/_3$ and $^1/_5$ of the population notice this smell. It seems probable that sulfur-containing amino acids such as methionine are broken down in the body to other sulfur-based compounds. These, like many sulfur-containing compounds, have an unpleasant smell. (Hydrogen sulfide, H_2S, is the compound that causes the smell of bad eggs.)

The reason that not everyone can smell 'aspragus-pee' is the subject of debate. Scientists remain divided on why people have different urinary responses to eating asparagus. One camp thinks only about half of the population have a gene enabling them to break down the sulfur-containing amino acids in asparagus into their smellier components. Others think that everyone digests asparagus the same way, but less than half of us have a gene that enables us to smell the specific compounds formed in the digestion of asparagus. More information on this topic can be found at the following web sites (amongst others):
http://www.boston.com/globe/search/stories/health/how_and_why/060694.htm
http://www.studentbmj.com/back_issues/0800/education/277.html
http://my.webmd.com/content/article/43/1671_51089
(accessed January 2005)

The questions on the student sheet are reproduced below, followed by the answers.

Questions

1. The amino acid methionine has the formula

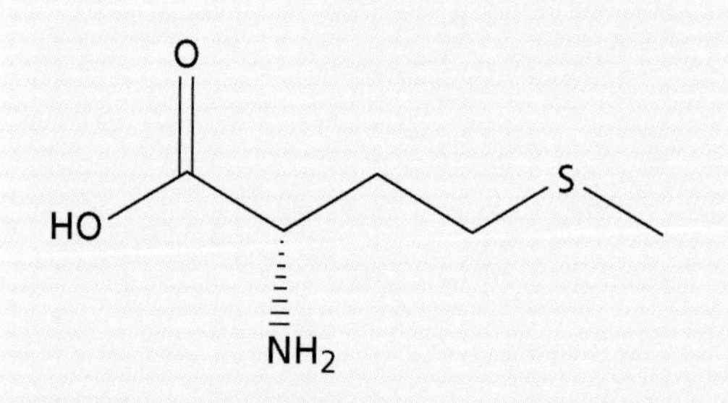

Perhaps surprisingly the amino acid asparagine contains no sulfur. Find out the names and formulae of the two other naturally-occurring amino acids that contain sulfur.

2. This question is about 'asparagus-pee'.

It has long been known that after eating asparagus, many people can detect an odd odour in their urine. (Whether or not a person can actually smell it is apparently genetic.)

In the 19th century it was thought that the compound responsible for the smell in so-called 'asparagus-pee' was methanethiol, CH_3SH.

In a more recent study, no methanethiol was detected in the vapour from asparagus-pee but two other sulfur-containing compounds, **A** and **B**, were identified by mass spectrometry. The M^+ ion for compound **A** is at *m/e* = 102; for compound **B**, *m/e* = 150.

Asparagus

For comparative purposes, compounds **A** and **B** were also prepared in the laboratory from propenoic acid by the following routes.

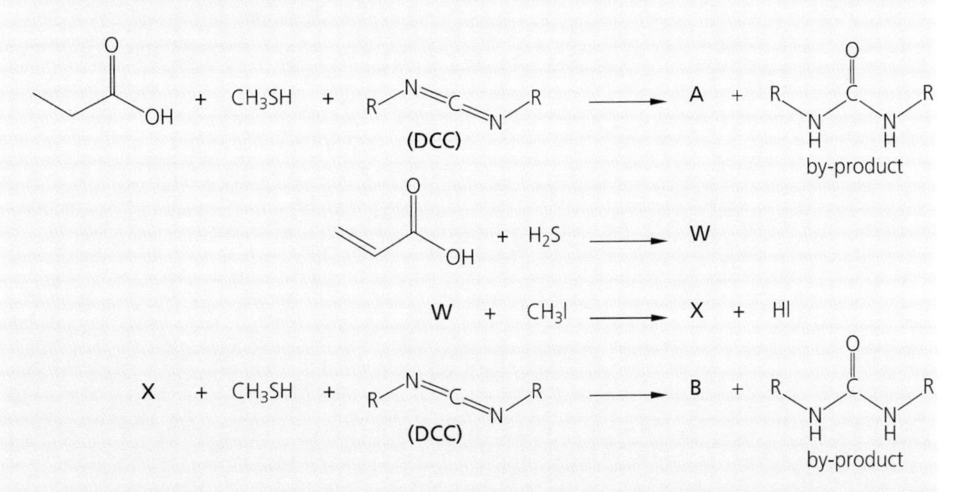

Compound **A** reacts with HBr to give **Y** and with Br_2 to give two isomers, **Z1** and **Z2**.

Y is not optically active and has two peaks for its M^+ ion at *m/e* = 182 and 184 (1:1 ratio).

Z1 and **Z2** are enantiomers (optical isomers) and their M^+ ions have *m/e* = 260, 262, and 264 (ratio 1:2:1).

(a) State the role of the reagent 'DCC': i) catalyst; ii) hydrating agent; iii) dehydrating agent; iv) hydrogenating agent or v) dehydrogenating agent.

(b) Suggest structures for compounds **A**, **B**, **W**, **X**, **Y**, **Z1**, and **Z2**.

(c) Both **A** and **B** may be hydrolysed in water, releasing methanethiol (hence the earlier suspicion of this compound). Write balanced equations for the hydrolysis reactions of both **A** and **B**.

[Naturally occurring bromine is a 1:1 mixture of ^{79}Br and ^{81}Br]

Answers

1. Cysteine

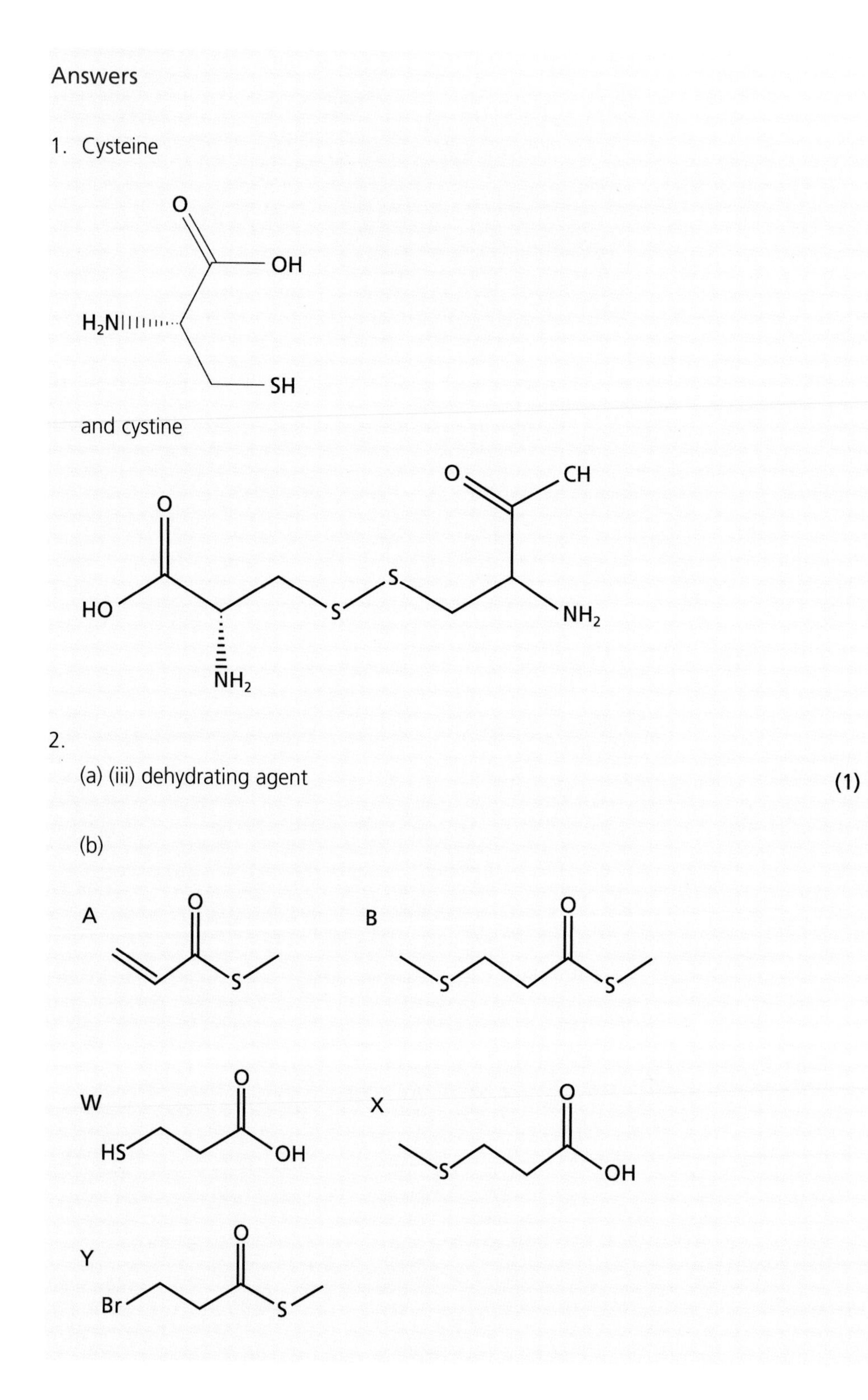

and cystine

2.

(a) (iii) dehydrating agent **(1)**

(b)

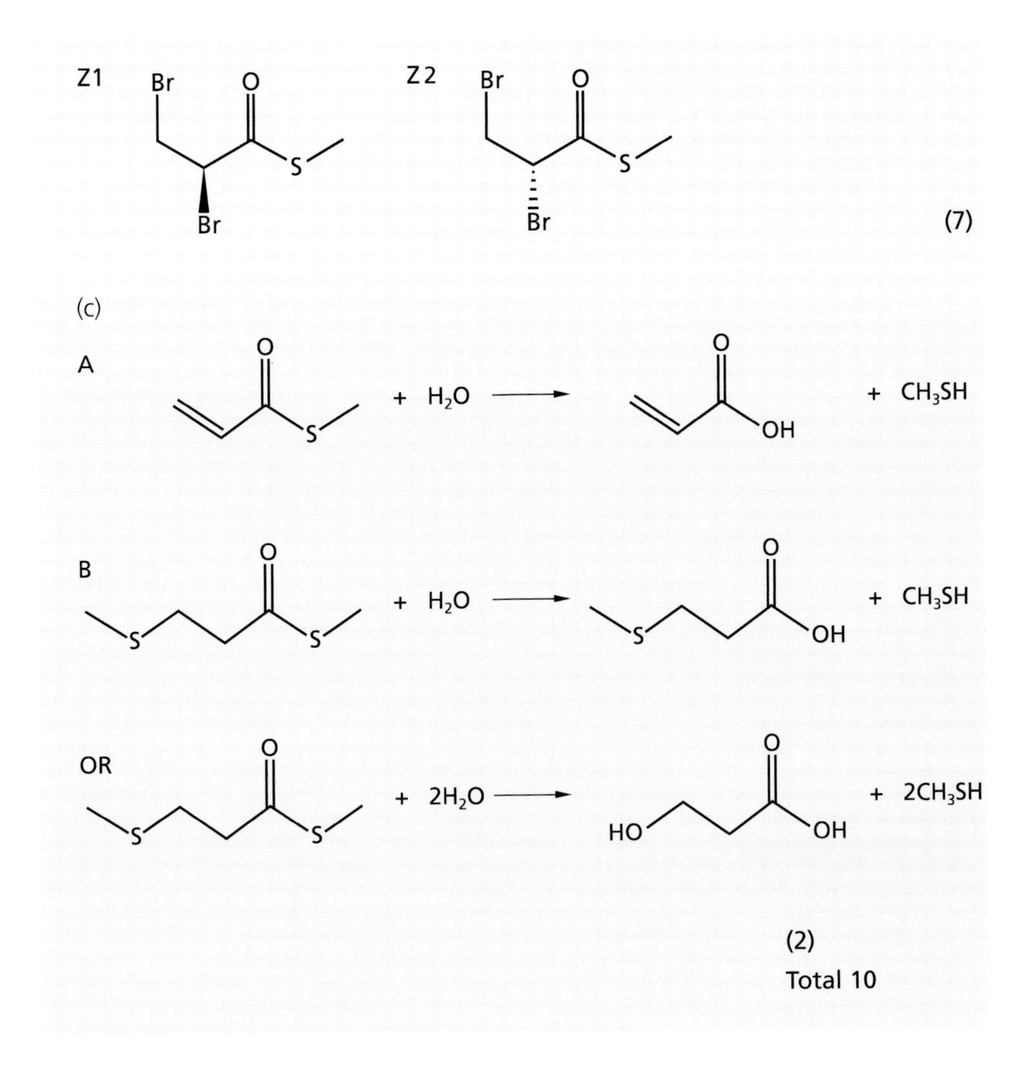

Teacher's notes:

....

....

....

....

....

....

....

....

....

....

....

Kitchen**Chemistry**

How hot are chilli peppers?

How hot are chilli peppers?

Learning objective

- Revision of organic structural formulae, isomerism, organic reactions and chromatography.

Level

Age: post-16.

Timing

About half an hour.

Description

A passage of reading about the chemistry of compounds responsible for the burning sensation caused by chilli peppers is presented, followed by questions that test the understanding of familiar chemistry in an unfamiliar context. Topics covered include organic structural formulae, isomerism, organic reactions and chromatography. The resource could be used for revision, as a homework exercise or in the case of teacher absence.

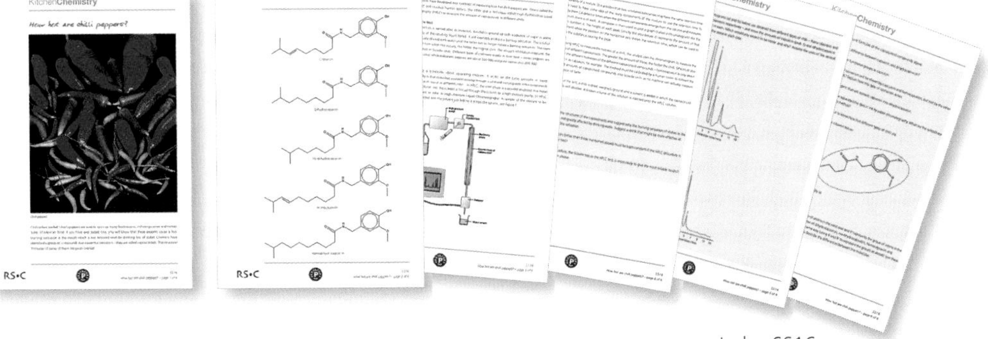

Index SS16

Teaching notes

This material may be used alone, but there are links with the material in ***The Chemistry of flavour*** (see page 77) and chillies are mentioned in ***Enzymes and jellies*** (see page 73). The passage from the student's material is reproduced below along with answers to the questions.

How hot are chilli peppers?

Chilli (often spelled 'chile') peppers are used to spice up many food recipes, including curries and various types of Mexican food. If you have ever tasted one, you will know that these peppers cause a hot, burning sensation in the mouth that is not removed even by drinking lots of water. Chemists have identified a group of compounds that cause this sensation – they are called capsaicinoids. The structural formulae of some of them are given overleaf.

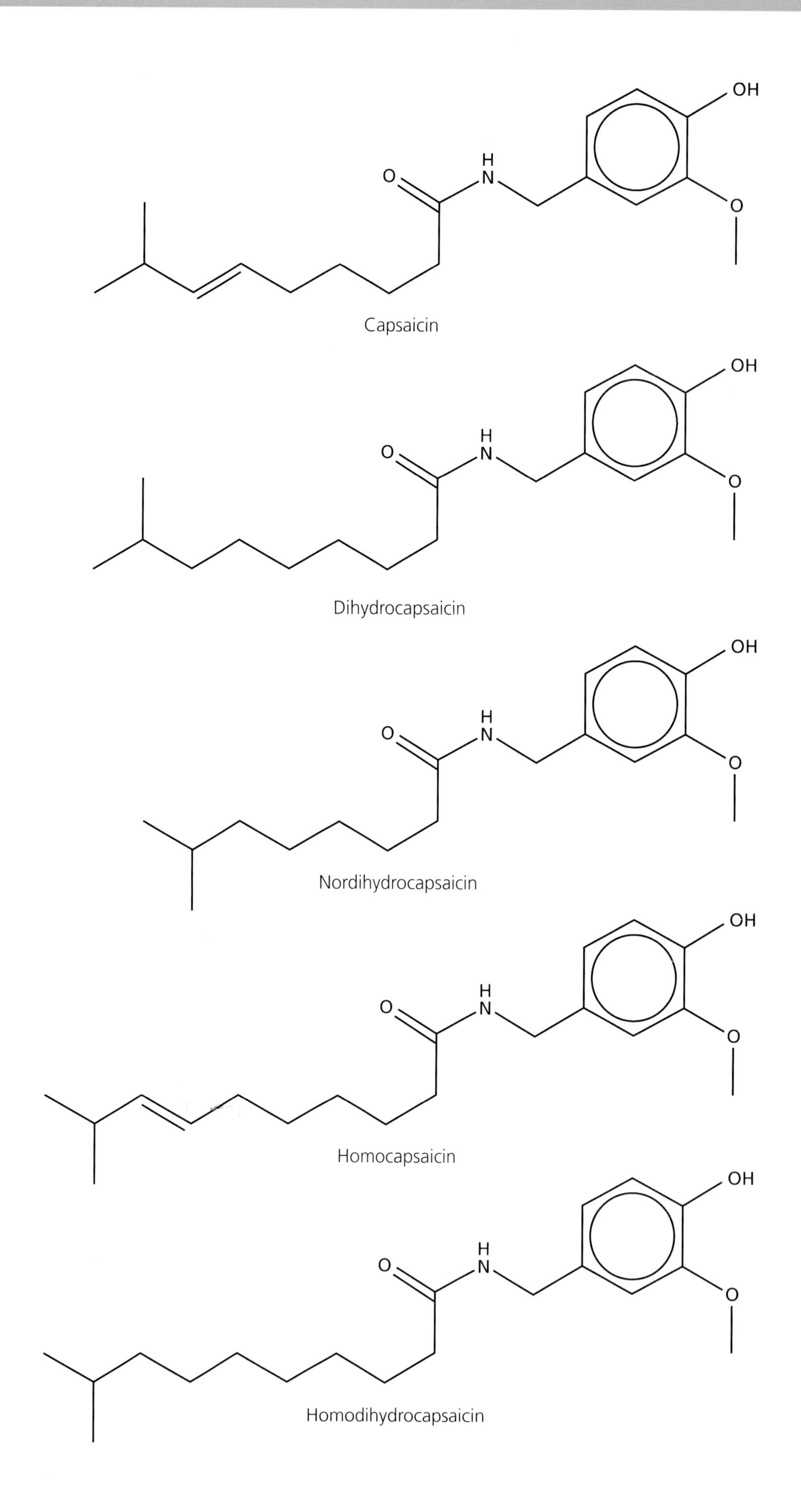
OH
H
N
O
O
Capsaicin
OH
H
N
O
O
Dihydrocapsaicin
OH
H
N
O
O
Nordihydrocapsaicin
OH
H
N
O
O
Homocapsaicin
OH
H
N
O
O
Homodihydrocapsaicin

Food scientists have developed two methods of measuring how hot chilli peppers are. One is called the Scoville test and involves human tasters, the other uses a technique called High Performance Liquid Chromatography (HPLC) to measure the amount of capsaicinoids in different chillies.

The Scoville test

In this test (which is named after its inventor), the chilli is ground up with a solution of sugar in water and a sample of the resulting liquid tasted. It will inevitably produce a burning sensation. The solution is then gradually diluted with water until the taster can no longer notice a burning sensation. The more dilute the solution when this occurs, the hotter the original chilli. The amount of dilution measures the heat of the chilli in Scoville units. Different types of chilli vary widely in their heat – sweet peppers are around 1000 units while Habanero peppers are about 100 000 and pure capsaicin is 350 000.

HPLC

This technique is essentially about separating mixtures. It works on the same principle as paper chromatography in that it involves a solvent moving through a solid (called the stationary phase) and carrying with it the components of a mixture, which travel at different rates. In HPLC, the stationary phase is a powder enclosed in a metal tube called a column and the solvent is forced through the column by a high pressure pump, so HPLC is sometimes said to refer to High *Pressure* Liquid Chromatography. A sample of the mixture to be separated is injected into the solvent just before it enters the column, see Figure 1.

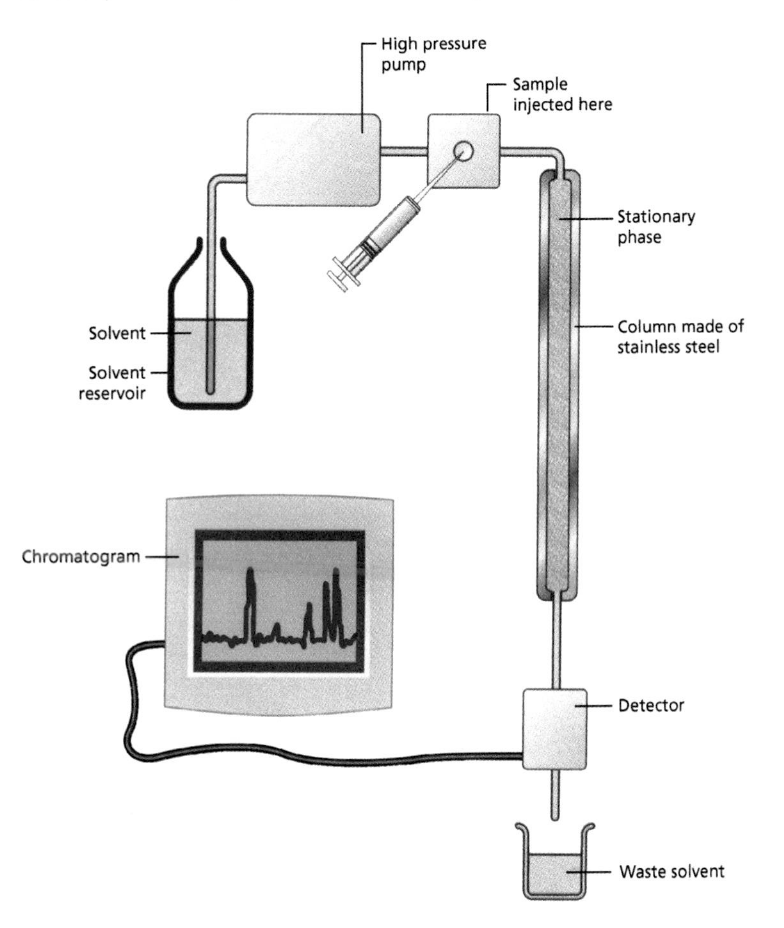

Figure 1 HPLC apparatus

The better a component forms intermolecular bonds with the liquid, the faster it moves through the column and the sooner it emerges from the other end. The time a component spends on the column is called the **retention time**. A particular substance will always have the same retention time provided that the solid phase and the solvent, as well as several other factors, are kept the same so the retention time can be used to identify components of a mixture. (It is possible that two unrelated substances may have the same retention time so we need to have some idea of the likely components of the mixture to use the retention time to identify them.) A detector times when the different components emerge from the column and measures how much there is of each. A computer is then used to plot a graph (called a chromatogram) like the ones shown in question 4. The height of each peak (strictly the area below it) represents the amount of that component present in the mixture while the position on the horizontal axis shows the retention time, which can be used to identify the substance causing the peak.

When using HPLC to measure the hotness of a chilli, the analyst uses the chromatogram to measure the amounts of different capsaicinoids present. The greater the amount of these, the hotter the chilli. She or he must also allow for the different hotnesses of the different capsaicinoid compounds – homocapsaicin is only about half as hot as capsaicin, for example. The method must be calibrated by a human taster to convert the measured amounts of capsaicinoid compounds into Scoville units as no machine can actually measure the sensation of taste.

To carry out the test, a chilli is dried, weighed, ground and a solvent is added in which the capsaicinoid compounds will dissolve. A known volume of this solution is injected onto the HPLC column.

Questions

1. Look at the structures of the capsaicinoids and suggest why the burning sensation of chillies in the mouth is not greatly affected by drinking water. Suggest a drink that might be more effective at reducing the sensation.

2. What factors (other than those mentioned above) must be kept constant if the HPLC procedure is to be a fair test?

3. Which procedure, the Scoville test or the HPLC test, is likely to give the most reliable results? Explain your choice.

4. HPLC chromatograms (a) and (b) below are obtained from different types of chilli – Patna oleoresin and Tezpur oleoresin respectively – and show the amounts of capsaicin (peak 1) and dihydrocapsaicin (peak 2) in each. Which would you expect to be hotter and why? Assume the units of the vertical axis are the same in each case.

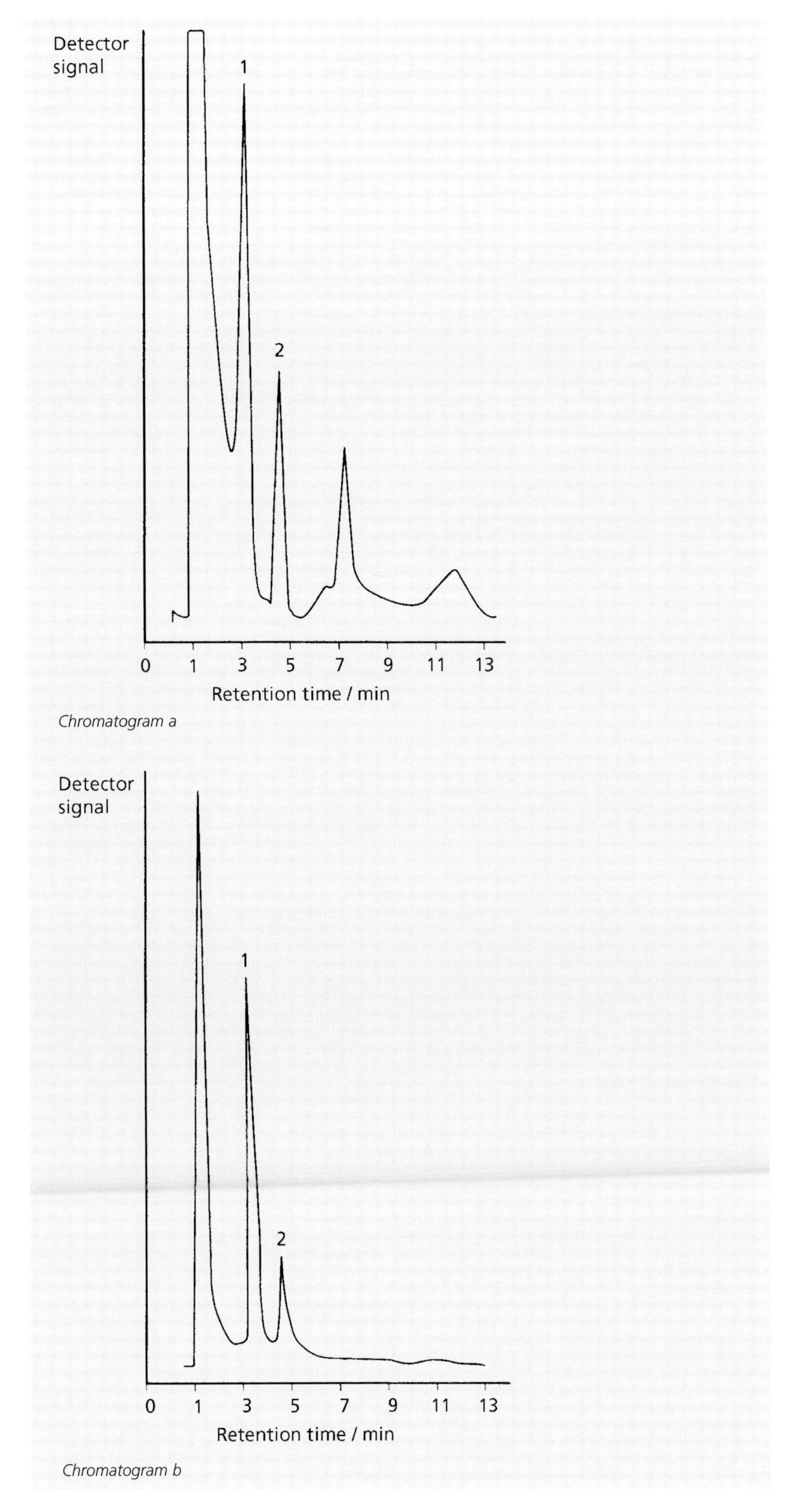

Chromatogram a

Chromatogram b

5. Look at the structural formulae of the capsaicinoid compounds above.

   (a) What is the difference between capsaicin and dihydrocapsaicin?

   (b) Name all five functional groups in capsaicin.

   (c) What type of isomerism will be shown by capcsacin and homocapsaicin, but not by the other three compounds? Explain how this type of isomerism arises.

   (d) Suggest a reagent that will convert capsaicin into dihydrocapsaicin.

6. You will probably have separated the dyes in ink by paper chromatography. What are the solid phase and the solvent in this method?

7. Suggest why it is useful to know how hot different types of chilli are.

8. Look at the formula for capsaicin below

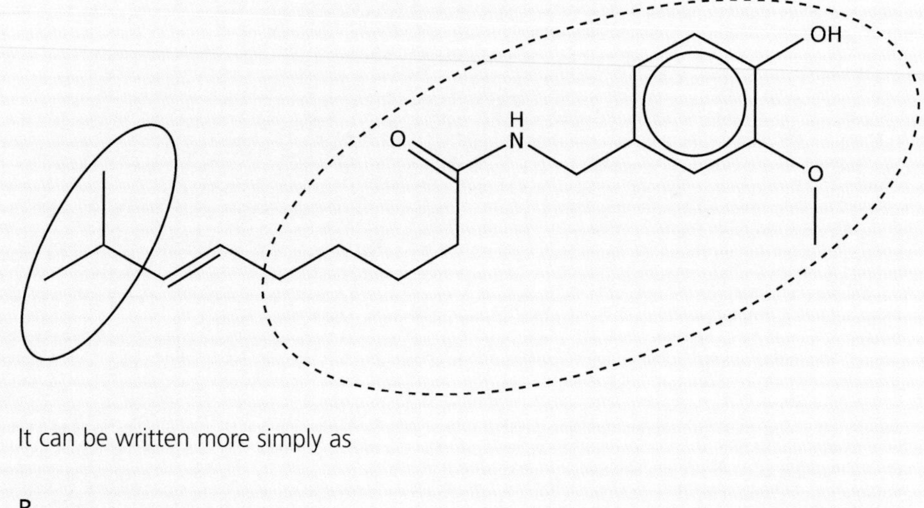

It can be written more simply as

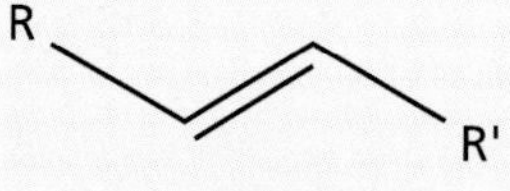

where R represents the group of atoms in the solid oval and R′ represents the group of atoms in the dotted oval. Draw the formulae of dihydrocapsaicin, nordihydrocapsaicin, homocapsaicin and homodihydrocapsaicin in the same way (using R and R′ to represent the groups as above). Use these representations to help you to describe the differences between the molecules.

Further information

The following websites give further information about chillies and how their pungency is measured. Many more relevant sites can be found by using a search engine.

http://www.chillipepperinstitute.org/Pungency.htm
http://www.fiery-foods.com/dave/assam_chilli1.html
http://www.bbqdan.com/bbq_docs/wilbur_scoville.html
(accessed January 2005)

Answers

1. The capsaicinoid molecules are relatively non-polar. They have an aromatic ring and a long hydrocarbon chain along with some polar groups. Thus they are likely to be relatively insoluble in water and not readily washed from the mouth by a drink of water. Milk might be a more effective drink as the relatively non-polar capsaicinoid molecules would be expected to dissolve in the fat globules contained in milk.

2. A standard mass of chilli must be used. The amount of solvent in which the chilli is dissolved must be kept the same. Pressure, temperature, flow rate and the length and diameter of the column can all affect retention times as well and must be kept constant.

3. The HPLC test should be more reliable as the Scoville test depends on human tasters who may differ from day to day and from one another.

4. Chilli (a) would be hotter because the peak heights for both capcsacin and homocapsaicin are higher in chromatogram (a) than in (b).

5. (a) The double bond in capsaicin has been hydrogenated in dihydrocapsaicin.

   (b) Aromatic ring, phenol, ether, amide and alkene.

   (c) *Cis-trans* isomerism. This is a result of the fact that there is no free rotation about a carbon-carbon double bond.

   (d) A reducing agent such as hydrogen with a nickel catalyst will reduce C=C but not C=O. Note that reagents such as sodium tetrahydridoborate(III) and lithium tetrahydridoaluminate(III) are unsuitable as they would reduce the C=O group.

6. The solid phase is the paper (strictly water bonded to the paper) and the liquid phase is water (or other solvent such as ethanol or propanone).

7. Chefs and food processing companies will need to know how hot chillies are in order to know how much to add to their recipes. Chefs may well do this by personal tasting but food processing companies may well prefer a more objective measurement.

8. The representations are:

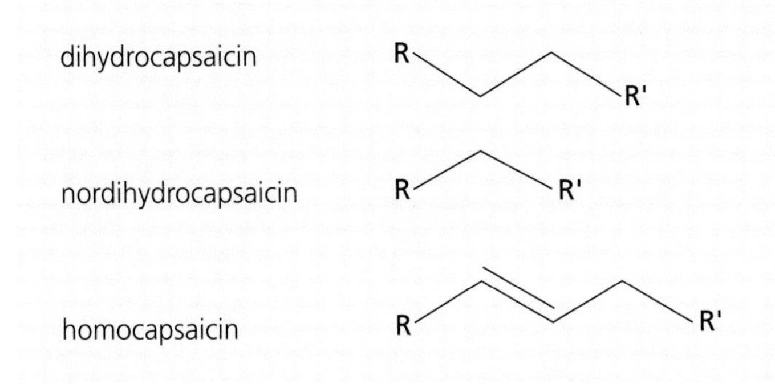

homodihydrocapsaicin R⁄\⁄\ R'

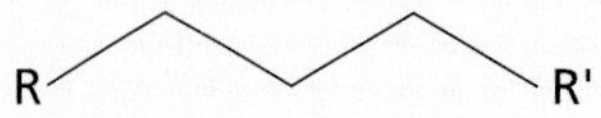

They show that capsaicin and dihydrocapsaicin differ only in that the former has a carbon-carbon double bond while the latter does not.

Nordihydrocapsaicin lacks the double bond and has a carbon chain one carbon atom shorter than that in capsaicin. Homocapsaicin and homodihydrocapsaicin both have a carbon chain one carbon atom longer than capsaicin. Homocapsaicin has a carbon=carbon double bond and Homodihydrocapsaicin does not.

Teacher's notes:

Teacher's notes: